XYLITOL

An International Symposium organised and sponsored by
Roche Products Limited and Xyrofin Limited
at the Europa Hotel, London, on 5th May, 1977

ORGANISING COMMITTEE

J. N. COUNSELL, P. J. MARTINDALE, D. J. ROBERTON
(*Roche Products Limited*)

XYLITOL

Edited by

J. N. COUNSELL

B.Sc., F.I.F.S.T., M.R.S.H., M.Inst.M.

Manager, Technology Centre, Roche Products Limited, Chemicals Division, Dunstable, UK

APPLIED SCIENCE PUBLISHERS LTD
LONDON

APPLIED SCIENCE PUBLISHERS LTD
RIPPLE ROAD, BARKING, ESSEX, ENGLAND

British Library Cataloguing in Publication Data

Xylitol.
1. Dental caries—Prevention—Congresses
2. Xylitol—Physiological effect—Congresses
I. Counsell, J N II. Roche Products Limited
III. Xyrofin Limited
617.6′01 RK331

ISBN 0-85334-786-7

WITH 10 TABLES AND 59 ILLUSTRATIONS

Printed in Great Britain by Galliard (Printers) Ltd, Great Yarmouth

Acknowledgements

Among the many who contributed to the success of the symposium, three deserve the editor's warmest personal thanks: Professor Gey for his help in deciding which speakers to ask and for being able to persuade them to take part, Mr Marshall Midda for his constant encouragement and for making coherent sense of the discussions and Miss Pauline Martindale, upon whose shoulders fell most of the organisation and the seemingly endless tasks of preparing the preprints, and typing and retyping the manuscript of this volume.

Foreword

Within the last decade the dental profession, realising that there is a severe manpower shortage to deal with established disease, has started to place more stress on preventive techniques. Research at laboratory and clinical levels has shown that periodontal disease can be prevented and controlled by the daily removal of dental plaque using a correctly designed brush in the correct way and interdental cleansing with floss.

However, caries is still the second most common disease in the world and costs the taxpayers of England and Wales a minimum of £200 million a year to repair the damage. We know that in any one year more working days are lost owing to dental disease than all of the strikes put together. Nobody can put a value judgement on the disfigurement and pain that a child may have to endure.

The interesting part about dental caries is that, since 1890, the cause of the disease is still thought to be the same. Every school child knows that sugar plus bacteria causes acid and that the acid eats your tooth away. For there to be decay, there has to be a tooth, there has to be bacterial plaque and there has to be a substrate on which the bacteria can work. By removing any one of these factors, we can eliminate decay. With better oral hygiene and thus reduction of plaque will decay be controlled?

Some authorities have said that a clean tooth does not really matter because it can still decay, but recent research projects have shown that good oral hygiene can, in fact, reduce the incidence of caries. For example, the National Child Survey and the work of Linden and Axellson in Sweden show some evidence of this. Recently, at Bristol, Andlaw and Tucker in the MFP toothpaste trials showed that there was a statistically significant reduction in the caries rate of children who had good oral hygiene. We have a long way to go, however, in a population where the average purchases are 0·7 toothbrushes per head of population per annum. As a nation, it would seem that we still spend more money on our armpits than we do on our mouths.

The use of fluoride to protect the tooth surface is well documented. Water fluoridation is a primary preventive measure and has a cost benefit to the nation. There were many objections on health grounds—it was said to have caused everything from acne to impotence, although the last one was difficult to disprove. Now the Royal College of Physicians Report has cleared the health hazard the argument has reverted to the problem of an individual's rights. However, the alternative to water fluoridation, such as tablets, milk and salt, do not have the same effects at a community health level, since the socio-economic groups that most need the benefit of fluoride are the ones least likely to take advantage of these methods.

Fissure sealants at this time are unlikely to be accepted as a public health measure until they can be applied more easily and at a cost benefit level. Can we, therefore, protect the tooth by dietary control?

By eliminating the sugars that are known to be associated with decay, a reduction must take place. We certainly know that there is a direct relationship between sucrose intake and the DMF rate. In this country, and other countries such as Australia, Holland, Switzerland and Iceland that have some of the highest sugar intakes in the world, caries is a major problem. Trying to change the behaviour of a nation is probably the most difficult task of all.

In the past we dentists have frustrated our patients. We have said—'don't eat sweets, brush your teeth after every meal and come and see us every six months'—and patients have obeyed these rules and still lost their teeth. Perhaps the time has come to change these rules of dental health.

The British Dental Health Foundation recommends:

1. remove plaque from each surface of each tooth once a day;
2. use sugar sensibly;
3. fortify with fluoride;
4. visit your dentist regularly for a health check.

Man has a sweet tooth. Perhaps we can control this habit in spite of ourselves through the use of sugar substitutes—especially when used between meals.

The results of the now famous Turku studies gave an indication that some form of dietary control of caries may eventually be possible. In the aftermath of the studies, many questions have been raised. Is it really necessary to replace all sucrose in the diet? Is xylitol completely safe? Does it really cause diarrhoea and, if so, is it a short-term effect? These questions and many others have led to a widely diversified research effort involving workers in many countries.

By sponsoring this symposium, Xyrofin Limited and Roche Products Limited hoped to achieve a number of objectives. A need was seen for bringing together the leaders of the various European research activities in order to review the present state of knowledge of xylitol. The stimulus of hearing the lecturers summarise their work on different fronts was expected to lead to a valuable discussion, especially enabling a rare confrontation to take place between the food technologist (particularly the confectioner) and the dental practitioner.

The papers presented on the Food Technology, Biochemistry and Dental Science aspects cover the subject matter well. I believe that the symposium achieved the objectives which the sponsors and organisers set and that this book will indeed provide a valuable reference point in the history of xylitol.

MARSHALL MIDDA
Director of the School for Dental Hygiene
University of Bristol

Contents

Principal Participants

Mr C. Aminoff
Managing Director, Xyrofin Limited, Baar, Switzerland.

Professor Dr K. H. Bässler
Professor of Physiological Chemistry, Johannes Gutenberg University, Mainz, W. Germany.

Professor A. E. Bender
Professor of Nutrition, Queen Elizabeth College, University of London, London, UK.

Dr T. E. Doty
Scientific Research Manager, Xyrofin Limited, Baar, Switzerland.

Professor Dr H. Förster
Klinikum der Johann Wolfgang Goethe University, Frankfurt-am-Main, W. Germany.

Professor Dr G. Frostell
Faculty of Odontology, Lund University, Malmö, Sweden.

Professor Dr F. Gehring
Professor of Experimental Dentistry, Bayerische Julius-Maximillians-University, Würzburg, W. Germany.

PROFESSOR DR F. GEY
Director of Research, Xyrofin Limited, Baar, Switzerland.

DR F. KRACHER
Food Technologist, Givaudan Dübendorf Limited, Dübendorf, Switzerland.

PROFESSOR DR K. K. MÄKINEN
Professor of Biochemistry, Institute of Dentistry, University of Turku, Turku, Finland.

PROFESSOR DR A. SCHEININ
Professor of Cariology, Institute of Dentistry, University of Turku, Turku, Finland.

MR E. VANNINEN
Marketing Manager, Xyrofin Limited, Baar, Switzerland.

MR F. VOIROL
Food Technologist, Xyrofin Limited, Baar, Switzerland.

CHAPTER 1

The Occurrence, Manufacture and Properties of Xylitol

C. Aminoff, E. Vanninen and T. E. Doty
(*Xyrofin Limited, Baar, Switzerland*)

THE NATURAL OCCURRENCE OF XYLITOL

Xylitol is a pentahydric sugar alcohol, or pentitol, which occurs widely in nature in many fruits, berries and vegetables (Table 1). Yellow plums or greengages perhaps have the highest content—almost 1 % on a dry solids basis. Xylitol has thus always been a minor component of the human diet. It would, however, be very uneconomical to extract pure xylitol from such sources due to their own high cost and the relatively low concentrations of their xylitol content.

RAW MATERIAL SOURCES FOR XYLITOL

Glucose and xylose, being the basic building blocks of cellulose and hemi-cellulose which form the structural material of all plants, are the most widely occurring of all sugars. While cellulose consists entirely of glucose polymers, the hemi-cellulose of many plants consists primarily of xylans or xylose polymers. Included among such xylan-rich plant materials are, for example, various hard woods, oat and cotton-seed hulls, corn (maize) cobs, sugar-cane

TABLE 1
Occurrence of Xylitol in Various Fruits and Vegetables (Washüttl, Riederer and Bancher, 1973)

Product	*Xylitol in mg/100 g dry substance*
Banana (*Musa sapientum* L.)	21
Raspberry (*Rubus idaeus* L.)	268
Strawberry (*Fragaria* var.)	362
Yellow plum (*Prunus domestica* subspecies *italia*)	935
Carrot (*Daucus carota* L.)	86·5
Endive (*Cichorium endivia* L.)	258
Onion (*Allium cepa* L.)	89
Lettuce (*Lactuca sativa*)	131
Cauliflower (*Brassica oleracea* L. var. *botrytis*)	300
Spinach (*Spinacia oleracea* L.)	107
Aubergine (*Solanum melongena* L.)	180
Yellow boletus mushroom (*Boletus edulis* Bull.)	128
Brewer's yeast	4·5
Chestnut (*Castanea vesca*)	14

bagasse, straw and nut shells, to mention only a few. The xylan content of these materials amounts to around 20–30 % of the dry substance.

The fact that xylan is more easily hydrolysed than is cellulose offers the technical possibility for xylose and xylitol production. With current technology it is possible to recover from 50–60 % of the raw-material xylan as xylitol.

THE LARGE-SCALE PRODUCTION OF XYLITOL

The four main steps in the production of xylitol (Fig. 1) are acid hydrolysis of the plant material; purification of the hydrolysate to either a pure xylose solution or a pure crystalline xylose; hydrogenation of the xylose to xylitol; and crystallisation of the xylitol. The current process employed by Xyrofin Ltd is based on birch wood chips. The residue of the hydrolysis which consists of cellulose and lignin can be further utilised by the wood-processing industry in the production of chipboard, cellulose or cardboard.

The hydrolysate undergoes a complicated process of purification to remove the unwanted foreign substances which have also been dissolved during the hydrolysis and which provide another significant by-product for anti-ketotic and other livestock feed

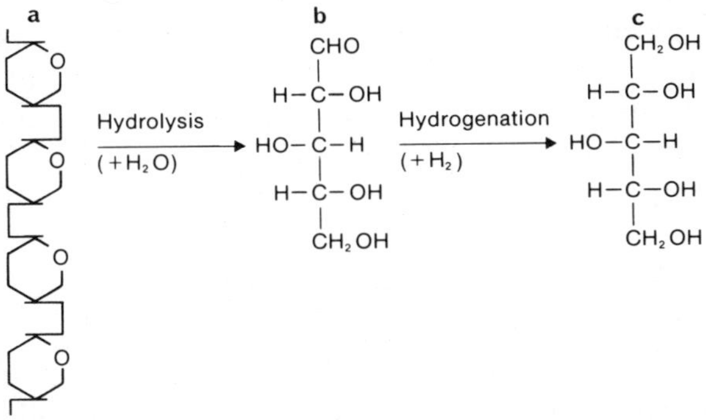

FIG. 1. Principle of xylitol production. a, Xylan $(C_5H_8O_4)n$, $n \sim 200$. b, D-Xylose, $C_5H_{10}O_5$. c, Xylitol, $C_5H_{12}O_5$.

supplements. Ion-exchange chromatographic methods developed by the Finnish Sugar Co. Ltd are employed for this purpose. After hydrogenation of the pure xylose solution xylitol is removed from the water solution by crystallisation.

The production of xylitol is an investment-intensive manufacturing process and the production costs are capacity dependent. Thus, to a large extent, the more that can be produced, the lower the costs and final unit price will be. A considerable know-how lead and research investment are also embodied in the already existing, large-scale plant which may be difficult to overcome for the newcomer who wishes to enter the field of xylitol production at a later date.

Regarding alternative methods of production, small amounts of xylitol can be produced or synthesised on a laboratory scale starting from glucose, using microbiological or synthetic methods. For the foreseeable future, however, such routes are out of the question for mass production of xylitol. The expenditure would be too great and the yields too small.

THE RATIONALE FOR XYLITOL

Principal Physical-Chemical Properties

Primary interest in xylitol centres on its properties and potential uses as a sugar substitute. First of all it is sweet; at room temperature it is equally as sweet as sucrose and therefore approximately twice as sweet as sorbitol and nearly three times as sweet as mannitol. In addition, when eaten in solid or crystalline form it gives a pleasant cool or fresh sensation due to its high negative heat of solution.

The calorific content of xylitol is also the same as that of sucrose (and other nutritive sweeteners), that is, approximately 4 kcal/g or 17 kJ/g. Thus, relative to level of sweetness, xylitol has only half the calories of sorbitol and one-third the calories of mannitol.

Figures 2 and 3 and Tables 2–4 present a comparison of some of the more important physical properties of xylitol and sucrose. The principal significance of these properties from the point of view of the food technologist is that, using xylitol in place of sucrose, it is

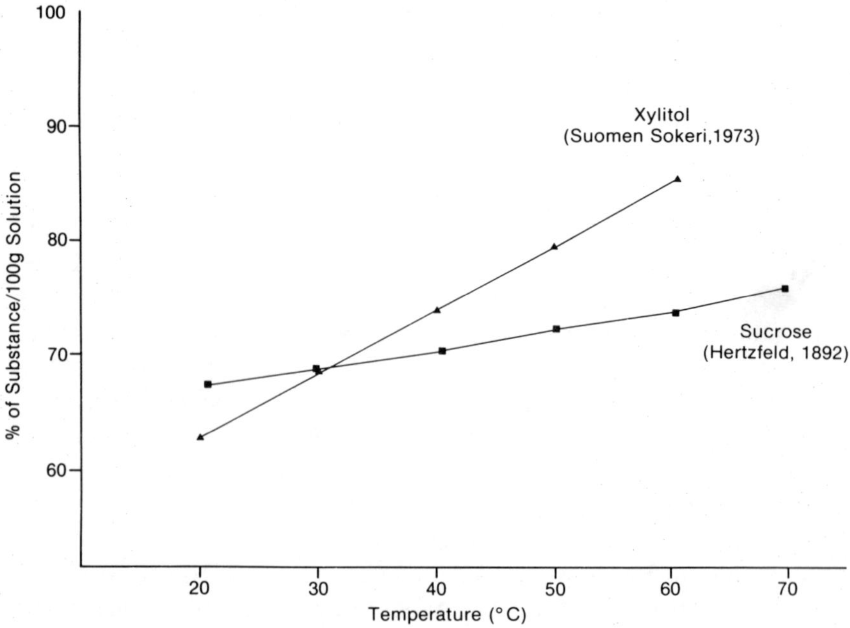

FIG. 2. Solubility in water of xylitol and sucrose.

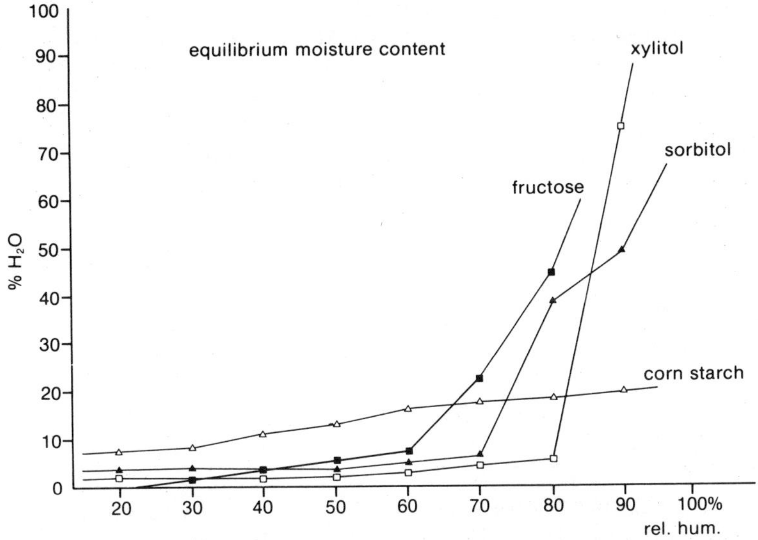

FIG. 3. Absorption isotherms for crystalline carbohydrates (Kammerer, 1972).

possible to produce chewing gums, confections and other sweet snacks, using essentially familiar formulations and methods to achieve final products which are qualitatively superior to their traditional sucrose counterparts (Voirol, 1977; Kracher, 1977).

Physiological Properties

In addition to its natural occurrence in foods, xylitol is also a normal metabolic intermediate in mammalian carbohydrate metabolism with an endogenous production and further utilisation of some 5–15 g daily in the average adult human. Total metabolic capacity for exogenous xylitol is even greater, i.e. some 600 g per day in parenteral administration (Bässler, 1977; Förster, 1977).

Slow absorption, and entry into metabolic pathways independently of insulin and without causing rapid fluctuation of blood glucose levels support the use of xylitol as a diabetic sweetener while extensive clinical and laboratory experience has shown the only untoward side effect of xylitol to be the possibility of an unpleasant but harmless osmotic diarrhoea after relatively large (20–30 g) initial oral doses in unadapted subjects (Bässler, 1977; Förster, 1977).

TABLE 2
Hygroscopicity and Equilibrium Moisture Content (EMC) (Seth, 1973)

Temperature (°C)	*Relative humidity (%)*	*Moisture absorption (%) of xylitol with storage* 1 *day*	2 *days*	6 *days*
23	76	0·04	0·118	0·308
	86	3·8	7·64	25·20
	92	21·7	37·7	92·34
			4 *days*	
37	45		0·048	
	55		0·054	
	63		0·054	

Perhaps the most significant characteristic of xylitol, however, is the fact that it is not utilised by the acid-producing, cariogenic bacteria of the human oral cavity. Xylitol is thus the first naturally occurring. completely physiological, non-acidogenic sweetener which has been demonstrated by comprehensive, long-term clinical trials to be non-cariogenic (Frostell, 1977, Gehring, 1977, Scheinin, 1977, Mäkinen, 1977). This fact clearly indicates the potential rôle of xylitol as an effective weapon in the fight against dental caries in addition to the traditional ones of prudent nutrition, conscientious

TABLE 3
Viscosity of Xylitol Solution in Centipoises (cP) (Kammerer, 1972)

g Xylitol[a] *per* 100 g *solution*	*Temperature (°C)* 20	30	40	50	60	70	80	90
10	1·225	0·969	0·819	0·638	0·562	0·508	0·426	0·388
15	1·491	1·133	0·934	0·761	0·653	0·550	0·480	0·426
20	1·798	1·376	1·112	0·874	0·782	0·621	0·535	0·470
25	2·180	1·656	1·332	1·060	0·843	0·717	0·613	0·537
30	2·714	2·103	1·574	1·245	1·003	0·840	0·704	0·610
35	3·529	2·516	1·915	1·497	1·214	1·032	0·844	0·709

[a] The xylitol was dried in a vacuum for 4 h at 55 °C before use.

TABLE 4
Comparison of Physical Properties of Xylitol and Sucrose (Based on Kracher, 1975)

Property		*Xylitol*	*Sucrose*
Melting point (°C)		93–94·5	179–186
Boiling point (°C/760 mmHg)		216	decomposes
Density (bulk density) (15 °C)		1·50	1·59
Caloric value (cal/g)		4·06	4·06
Fermentation		—	+ +
Caramelisation, browning		—	+ +
Moisture absorption (%) (4 days, 20–22 °C)			
60 % relative humidity		0·05	0·02
92 % relative humidity		90	4
Density (specific gravity) of aqueous solution $\left(\frac{20\,°C}{4\,°C}\right)$			
	10 %	1·03	1·04
	20 %	1·07	1·08
	40 %	1·15	1·18
	60 %	1·23	1·29
Heat of solution (cal/g), endothermic		36·61	4·34
Optical rotation, $[\alpha]_D^{20}$		—	+66·5
Viscosity (cP) (20 °C)			
	10 %	1·23	1·31
	20 %	1·67	2·03
	40 %	4·18	6·17
	50 %	8·04	16·42
	60 %	20·63	58·50
Index of refraction (25 °C)			
	10 %	1·3471	1·3475
	20 %	1·3620	1·3634
	30 %	1·3779	1·3806
	40 %	1·3951	1·3991
	50 %	1·4132	1·4194

dental hygiene, and regular and proper dental inspection and treatment.

XYLITOL IN PERSPECTIVE

The production and technological realities dictate that the price of xylitol will never fall to the level of common sugar. Its special characteristics, in particular its dental aspects must, therefore—and in our opinion—*can*, justify a higher price for xylitol. At any rate, xylitol cannot realistically be considered as a competitive threat to common sugar. First, the dental trials at the University of Turku demonstrated convincingly that a complete substitution of xylitol for sucrose in the diet is not necessary to bring dental caries under control. On the contrary, it was shown that a comprehensive replacement of sucrose in solid forms, which are eaten between meals and tend to dwell in the mouth, for extended periods of time, with a xylitol-sweetened counterpart was just as effective as total substitution. The amount of xylitol involved in this between-meal sweets substitution was in the neighbourhood of 5–10 g per day. This amount may account for the major portion of between-meal sweets, the forms in which sucrose is most harmful to dental health, but it is insignificant compared to the total sucrose consumption which commonly exceeds 100 g per day in most developed countries and most of which is consumed in conjunction with regular meals. In addition, compared to an annual world sucrose production of some 80 million tons, the total world xylitol production at present is only about 6000 tons per annum. Thus, while a substantial increase in xylitol production is, of course, to be expected, it will certainly not threaten the position of sucrose or the valid rôle of sucrose as one of the cheapest major sources of calories, an excellent preservative and as an enhancer of the palatability of other nutritious foods. More probably the effect of xylitol in this regard will be just the opposite.

REFERENCES

Bässler, K. H. (1977) Biochemistry of xylitol. This volume, p. 35.

Förster, H. (1977) Tolerance in the human. Adults and children. This volume, p. 43.

Frostell, G. (1977) Behaviour of xylitol in isolated plaque. This volume, p. 133.

Gehring, F. (1977) Cariogenic bacteria. This volume, p. 111.

Hertzfeld, A. (1892) *Z. Ver. Rübenzuckerind*, **42,** 181.

Kammerer, F. X. (1972) *Kak. + Zucker*, **24,** 5, 184.

Kracher, F. (1975) *Kak. + Zucker*, **27,** 3, 68 and **27,** 4, 108.
Kracher, F. (1977) Some aspects of food technology. This volume, p. 21.
Mäkinen, K. K. (1977) Biochemical findings in exocrine secretions in relation to peroral administration of xylitol. This volume, p. 67.
Scheinin, A. (1977) Incidence of clinical and experimental caries in relation to microbiological and biochemical xylitol-induced effects. This volume, p. 139.
Seth, P. (1973) Hoffman-la-Roche, Basle, cited by Kracher, F. (1975).
Suomen Sokeri. (1973) Unpublished data.
Voirol, F. (1977) The value of xylitol as an ingredient in confectionery. This volume, p. 11.
Washüttl, S., Riederer, P. & Bancher, E. (1973) Qualitative and quantitative study of sugar alcohols in several foods. *J. Food Sci.*, **38,** 1262.

CHAPTER 2

The Value of Xylitol as an Ingredient in Confectionery

F. VOIROL

(*Xyrofin Limited, Baar, Switzerland*)

PROPERTIES AND USE IN THE FOOD INDUSTRY

It is self-evident that Finland, the country with the largest production site and the most significant xylitol studies, would be where the first xylitol-containing product appeared on the market. The first manufacturers were those firms who had participated in the two-year Turku study by providing xylitol-containing foods. One producer of sweets simply used the brand-name 'Xylitol' for his new products, abstaining from any reference to the specific properties of the sugar replacer. As a result of the national pioneering efforts in this field, xylitol has become a well-known term to wide circles of population in Finland.

This paper describes some of the possibilities in xylitol food application with the resulting technological problems and special properties of these new products. Part of these recommendations are based on experiences that have not yet been evaluated commercially. However, the breakthrough for the application of xylitol in foods can now be seen in various countries.

HOUSEHOLD SUGAR REPLACER

Xylitol is easy to press into cubes or fill into sachets. When sweetening warm beverages, the temperature dependence of xylitol sweetness, being different from that of sugar, must be considered. Thus, to sweeten a cup of coffee, 13 g of xylitol will be needed to replace 10 g of sugar for no difference by triangle test.

In most industrialised countries, the percentage of sufferers from diabetes is rising. Diabetics, like healthy people, have a natural liking for sweets. For these people, the availability of a natural sugar substitute which tastes good and can be used similarly in the kitchen is true progress. In contrast to sugar, reasonable doses of xylitol do not stress the organism of the diabetic. Nevertheless, its calories must be accounted for in the allowable daily intake (so-called 'bread units') determined by the physician. However, it must be stressed that xylitol is not a low-calorie sweetener but a natural, sweet carbohydrate having the same calorific value as sugar.

PRESERVES, KETCHUP, CONDENSED MILK

The preservative properties of xylitol can be put to work in the preparation of the 'heavy syrup' used in canning fruit. A more significant use, however, is the replacement of sugar in highly sweetened products like ketchup or sweetened condensed milk. Both foods can spoil during manufacturing as well as during storage by caramelisation. The well-known discoloration, or browning, is a chemical reaction between sugar and proteins, accelerated by higher temperatures. Tomato ketchup contains 50% sugar in the dry matter. When replacing the latter with xylitol, a ketchup is obtained which will keep its red colour longer and which is less prone to spoil even after the bottle has been opened. The same applies to sweetened condensed milk. Such applications make practical use of the high chemical stability of xylitol. The pentahydric sugar alcohol has no aldehyde or keto groups and therefore, unlike other sugars, does not cause non-enzymatic brownings, the so-called Maillard reactions.

In jams, jellies and marmalades, xylitol may be used up to a concentration of 40% in place of sugar. At higher concentrations, the xylitol will crystallise during cooling of the product and during storage. The advantage of such products, besides their suitability for diabetics, is the possibility of having a high fruit to sweetener ratio. Jams containing a high percentage of fruit solids and less sugar are increasingly demanded by the consumer. As sugar in jams serves both as a sweetener and as a preserving agent, the addition of an *artificial* preservative becomes necessary in low-sugar jams. Xylitol, on the other hand, does not sustain the growth of moulds, yeasts or other microorganisms, even in low concentration. Thus, high-fruit/low-xylitol jams are possible without the addition of artificial preserving agents.

CONFECTIONERY

Most products of this class contain, besides sugar (sucrose), glucose syrup as a technological additive to control the crystallisation characteristics of sucrose. Due to its cariogenic properties, glucose syrup cannot be used in xylitol confectionery. Research was therefore directed towards tooth-sparing or even non-cariogenic substances to influence the crystallisation behaviour of xylitol.

Boiled Sweets (Hard Candy)

The recipe and manufacturing procedure for preparing boiled sweets from xylitol are the subject of two patent applications, one of which describes a glass-clear product with the addition of natural polymers, the other one a pure but non-transparent xylitol product (Austrian Patent Application 71/76 0019, 1976; Swiss Patent Application 8633/76, 1976). Both procedures permit the manufacturing of non-cariogenic sweets having a pleasant cooling effect when eaten. Previously, 'sugarless' sweets were made from sorbitol or hydrogenated starch derivatives but, as both substances are less sweet than sugar, the products had to be 'sweetened-up' with cyclamate or saccharin. Xylitol, being as sweet as sugar, does not require any artificial sweeteners, and candies made with it are superior in taste to any 'sugarless' candies so far available. As xylitol does not form acids on the tooth surface, pure xylitol high-boilings will be the first truly non-cariogenic products of this kind. The storage properties of samples over one year were good and there was no tendency to pick up moisture in a normal climate (25 °C; 60 % relative humidity) in open packagings.

Toffees, Caramels, etc.

The confectionery laboratory of the Food Research Association at Leatherhead, UK has performed trials on the manufacture of toffees with xylitol. Samples of both fruit-flavoured and milk solids-containing types were produced. The final report of the Association recommends either the addition of maltodextrin or Lycasin®.

Maltodextrin has been shown to be fermented by Streptococci and is therefore unsuitable. The influence of the addition of sorbitol on the crystallisation properties of xylitol is under investigation.

Lycasin® is a hydrogenated starch derivative which favourably influences the crystallisation characteristics of xylitol. Of the available commercial types, 'Polysorb 80/85' was shown to be 'tooth-saving' in tests carried out at the Dental Institutes of the Universities of Malmö and Zürich. Combinations of xylitol and Lycasin will be examined further. Their application in toffees and caramels seems to be promising as they avoid both the addition of artificial sweeteners normally necessary in pure Lycasin sweets and the hardening of pure xylitol products due to the crystallisation. Furthermore, the price can be kept low by the addition of 20–40 % of the relatively inexpensive Lycasin. Lycasin is currently permitted in some European countries.

Chocolate

Like fruit sugar, xylitol can be used to make tasty chocolate products. No artificial sweeteners have to be added nor has any modification to be made to the recipe since xylitol has the same sweetness as sugar. Xylitol is used in powder form. The conching process does not offer any difficulties but the relative humidity of the atmosphere in the manufacturing locality should be kept under control. With relative humidities exceeding 85 %, a coarse 'sandy' texture was noted in the product after storage. This may have been caused by a kind of 'hydrate film' around the xylitol particles.

Xylitol chocolates are on the market in Finland and the Soviet Union. The Russian product contains 43 % xylitol and its label bears the advice: 'for diabetics, under medical supervision'.

COMPRESSED TABLETS

Xylitol ground to powder-sized particles may be pressed into tablets after flavouring. Optimum functioning of the tablet press may depend on the relative humidity of the air at the production site.

The Finnish market offers peppermint tablets under the name of 'Bonadent' or 'Koivusokeri' (birch sugar). They consist of 99·25 % pure xylitol. The rest is described as 'flavours and additives'. Both direct-compression and granulation techniques can be used (see Appendix, p. 18).

COATINGS, DRAGÉES

The process of coating with xylitol has been studied extensively. Compressed xylitol nuclei of 0·22 g were coated with a xylitol syrup containing approximately 2 % gum arabic at 60 °C using a warm air jet for drying between layers. Flavours and colours can easily be incorporated in the coating liquors. Xylitol-coated chewing gums can be found on the market and the coating of other centres, such as chocolate or toffees is presently being investigated.

XYLITOL CHEWING GUM

Before the rôle of sugar in tooth decay became known in the 1950s, the chewing of gum was considered to be beneficial for the teeth. As chewing gum contains 50–75 % sugar, industrial reaction to sugar discrimination was an attempt to replace sugar with less cariogenic sweeteners. Because sugar is both a sweetener and a body-giving filler in chewing gum, it is not possible to replace it with cyclamates or saccharin alone, both of which are much sweeter than sugar.

With the industrial availability of sorbitol and mannitol—sugar alcohols of about half the sweetness of sugar—the manufacture of 'tooth-sparing' chewing gum became commercially feasible. An undesirable after-taste is inherent in these products, because saccharin or cyclamates must be used to bring their sweetness up to 'sucrose standards'. Many of the sorbitol/mannitol gums on the market use the term 'sugarless' on the label, which is somewhat misleading because all polyalcohols, including xylitol, although more suitable for diabetics, have the same calorific value per gram as ordinary sugar.

Some researchers believe that the chewing of pure unsweetened gum base would be beneficial to the teeth. There are three possible mechanisms:

1. Stimulation of the secretion of saliva to help 'washing down' the tooth surface and increasing the concentration of electrolytes favouring protection of the enamel.
2. A minor mechanical abrasion of plaque.
3. Removal of debris from interdental surfaces.

However, a chewing gum made of pure gum base is neither technologically nor commercially feasible.

The availability of xylitol made it possible for the first time, to produce a high-quality chewing gum, which is absolutely non-cariogenic. The Turku studies have shown that the frequency of intake and the time of contact with the tooth surface are of prime importance if xylitol is to be effective as an anti-cariogenic sweetener. The slow release from gum base makes chewing gum an excellent vehicle for this purpose.

The surprising results of the Turku chewing-gum study induced an interest in xylitol among most major and smaller chewing-gum manufacturers. Laboratory trials have shown that glucose syrup in xylitol chewing gum can be replaced by glycerol, gum arabic, aqueous xylitol solutions and/or special gum bases. Industrial trials have confirmed that cariogenic additives can be avoided in chewing gum (Table 1).

TABLE 1
Xylitol Chewing Gum

Recipe	*A*	*B* (*Control*)
Gum base[a]	30%	30%
Xylitol solution in water (64%)	12%	—
Xylitol, powdered, grain size 50 nm	56%	—
Sucrose	—	64%
Glycerol	1%	1%
Flavouring	0·6–1·0%	0·6–1·0%
Water	—	4%

Other glucose substitutes used include gum arabic.

Procedure: Soften the gum base in a Z-blade mixer at 45 to 50 °C; stir xylitol solution into the mass; add the powdered xylitol in three portions; mix flavour into the mass after the second addition; and finally, add the glycerol.

[a] Use special base for sugarless chewing gum.
e.g. Yalaco gum base made by Gum Base Company, Via Nerviano 25, 1-20020 Lainate, Italy.

Manufacturing procedures differ only to an insignificant extent from conventional methods. Depending on the gum base used, the kneading operation can be controlled with an aqueous xylitol solution or with glycerol. The xylitol used must be in the form of a powder. The median particle size should correspond approximately to that of normal powdered sugar. Difficulties may occur when milling in conventional, uncooled grinders. The unusually low melting point of xylitol entails a danger of caking. Machinery suppliers offer special grinders suitable for this purpose and the possibility of supplying powdered xylitol is currently being investigated.

When making stick-gum on rollers, the use of talcum as a dusting agent is recommended. Powdered xylitol is unsuitable due to its increased hygroscopicity. Besides ribbons and extruder-formed gums, chewing gum coated with xylitol has been studied extensively. The use of gelatin in warm coating with xylitol syrup appears to be indispensable for obtaining a regular cover. It is also possible to coat chocolate or compressed tablets with xylitol.

BAKERY GOODS

The manufacturing of various baked products where sugar is replaced with xylitol is possible in principle. However, if a crust formation, caramelisation or non-enzymatic browning is required, the addition of a small percentage of fructose is necessary. As stated earlier, xylitol does not take part in such reactions.

THE FUTURE

No-one should even consider the possible future replacement of the entire annual world production of 80 million tons of sugar by xylitol. Sugar, after fat, is the cheapest source of energy in human nutrition and may well remain so. Health consciousness in the industrialised countries will, however, increase the demand for sugar replacers quickly. Obviously, certain requirements must be fulfilled by such substances. They should be 'physiological' substances (well tolerated, natural or nature-identical); they should have the same sweetness and taste similar to sugar (sucrose is the taste standard); they should be suitable for diabetics; they should not pose technological problems in food processing; and they should not cause caries. All these conditions are fulfilled by xylitol which makes it not only one of the few new discoveries in the field of foods but also a valuable alternative to sugar as well as other sugar replacers.

APPENDIX

Xylitol Tablets (Direct-Compression Method)

Ingredients	*mg/tablet*	
Xylitol (cryst.)	500	500
Stearic acid (powdered)	8	8
Flavour FRESCOFORT 60470-31[a]	—	4
Tablet weight (mg)	508	512

[a] Flavours supplied by Givaudan & Co. Ltd, CH-8600 Dübendorf, Switzerland.

Manufacturing Procedure

Pass stearic acid (powdered) through a 0·22-mm sieve and mix with xylitol (cryst.) and flavour which have been passed through a 1·5-mm sieve. Compress on a Manesty RD3 single-rotary machine, a Manesty F3 single-punch machine or a similar compression machine. The hardness of the biconvex tablets should not exceed 3–4 SCU† otherwise the tablets will show capping. To obtain a low friability and for tablets of a hardness of 12–20 SCU, tablets should be treated as follows: store the tablets for 1 h at a maximum of 95 °C. After cooling slowly to room temperature (20–25 °C), the tablets show the desired hardness of 12–20 SCU.

Tablet Characteristics

Tablet weight (mg)	508	512
Form	biconvex, round	biconvex, round
Diameter (mm)	10·0	10·0
Radius of curvature (mm)	8·0	8·0
Thickness (mm) approx.	6·3	6·3
Height of tablet edge (mm) approx.	3·0	3·0
Hardness (Heberlein) (SCU)	12–20	12–20
Friability at 300 revolutions (%)	0·03	0·03

Xylitol Tablets (Granulation Method)

Ingredients	*mg/tablet*
Stearic acid	5·32
Magnesium stearate	2·66
Xylitol powder	532·02
Tablet weight (mg)	540

Preparation of Xylitol Powder

1. Xylitol 1 000 g
2. Polyvinylpyrrolidone as 20 % w/w solution in isopropanol 35 g

† Strong-Cobb units which are measured on a Heberlein hardness tester.

Fitzmill xylitol: #00 plate, medium speed, hammers forward.
Granulate with item 2
Dry at 140 °C overnight.
Fitzmill: #1A plate, medium speed, knives forward.

Manufacturing Procedure

A. Mix items 1 and 2 together, pass through #30 mesh and remix.
B Mix lubricants with item 3 for about 3–5 min.
C Compress on B3B rotary press using $\frac{1}{2}$ in s.c. plain-face punches.

Tablet Characteristics

Flow	Very good
Friability (%)	0·57
Hardness (SCU)	12–15
Thickness (mm)	approx. 6·35

CHAPTER 3

Some Aspects of Food Technology

F. KRACHER
(*Givaudan Dübendorf Limited, Dübendorf, Switzerland*)

As a food ingredient, xylitol is a new substance. Food manufacturers have only been aware of it since the mid-1960s. After the positive results of the Turku study, interest in xylitol grew considerably and many manufacturers started experimenting with it. It was very soon found that xylitol has some outstanding properties: it gives no browning reactions; it has a sufficient sweetness of its own; it is chemically very inert (does not invert); and it can be heated to melting point with no adverse effect.

PROPERTIES

Browning Reactions

When reducing sugars, i.e. carbohydrates with a carbonyl group (fructose, dextrose, sucrose), are in contact with amino acids, browning occurs. This process is also known as the Maillard reaction. There are many examples with confectionery and other products, where browning is definitely undesirable. In all these cases xylitol is recommended.

Sweetness

The following sweetness scale (Sucrose = 1·0) indicates that with xylitol the ratio of sweetness to weight is very high. In these days when artificial sweeteners are being banned, this is certainly a positive property.

Fructose	1·2
Xylitol	1·0
Sucrose	1·0
Dextrose	0·7
Sorbitol	0·5
Mannitol	0·5
Maltose	0·4
Lactose	0·3
Starch hydrolysates	0·3–0·8

Inertness

Xylitol being a 'monosaccharide' cannot invert. It is a very positive advantage of xylitol that it can be used in sweets manufacture together with fruit acids without alteration. For instance, a xylitol-coated sweet with acid and an orange flavour would be a real novelty on the market.

Melting

Being a nearly inert substance, xylitol can be heated to melting point (95 °C) without any alteration. This means that to apply xylitol in liquid form into a mass, it is not necessary to dissolve it in water (with subsequent evaporation). This is a very positive advantage of xylitol. But unfortunately there is also another side to the picture. Up until now, xylitol has always been compared with sucrose. There are several reasons why this comparison leads to psychological attitudes which are less favourable towards xylitol. For example, from the technological point of view, man has known sugar for centuries, during which he has built up the whole existing technology of sucrose. In other words, the entire confectioner's technology is based mainly on sucrose. When he thinks of sweets, he thinks of sucrose, the familiar properties of sucrose and the familiar reactions. For instance, when a confectioner wants to prepare a highly concentrated sugar solution, he boils the sucrose–water mixture to a predetermined temperature. The temperature/concentration ratio can be found in existing tables. The confectioner knows that when he heats the mixture to 219 °F (104 °C), he obtains a sucrose solution of 65 %. During the heating process a minor change in the sucrose also takes place—it becomes 'softer'.

These conditions no longer apply if xylitol is used. To the traditional confectioner, xylitol is a novelty, an unknown substance, a rival to his livelihood. His subconscious tells him, 'Be careful!' And it prevents the confectioner's thought process leading away from the customary properties of sucrose. For this reason he will also boil a xylitol–water mixture, although it does not give the expected results, and confers no 'softening' at all on the xylitol.

Again, from the organoleptic point of view, in former times, when man was still a collector of plant materials, he had to rely exclusively on his sense of taste in order to differentiate between the edible and the unacceptable. Generally, sweet was edible, bitter was noxious—unknown or unfamiliar tastes indicated danger; unfamiliar taste led usually to rejection. We, in the 20th Century, have not completely lost this old cautiousness. For centuries we have grown up with sucrose. Therefore, as soon as we taste sweetness, we expect the familiar sweetness and the texture of sucrose. If we detect another tonality, a cooling effect or a 'harshness', an uncontrollable mechanism in our subconscious triggers suspicion and tells us to be careful, and that something may be wrong!

All these aspects are not favourable for xylitol. How, then, can they be turned into positive advantages? There is but one 'thing' of which we all have unfortunately so little: time. After a certain period of time, more and more children will become familiar with xylitol, for these children, customarisation with xylitol will be much easier than it is for us today. Time cannot be built in with xylitol. Fortunately, there are enough favourable aspects of xylitol. We simply have to avoid the comparison of xylitol with sucrose. Instead, we should accept xylitol as 'The New Sweetener' and products containing xylitol as 'New Products of the 21st Century', just as we have done with many other food innovations. The novelty of the products should be supported by new flavour combinations like fruit with a touch of mint/herbs/bitter components/hot components/ 'medical' components/nutty notes.

There are thus many possibilities with xylitol.

Discussion: Xylitol as a New Ingredient

Chairman—PROFESSOR A. E. BENDER

Mrs A. Robertson (Dental Clinic, Chapel Street, Chichester, UK): Xylitol at the moment is being used as a sweetener for syrups. Have you in fact investigated its use to replace 60 % sucrose syrups which are currently being used as a base for sweetened medicines? I'm thinking particularly of the type of medicines which are given by mothers to children at night, asthmatics, sedatives, those types of sweeteners.

F. Voirol (Xyrofin Ltd, Baar, Switzerland): I believe there are two cough syrup products on the Finnish market. The use of xylitol to replace sucrose in the heavy syrup of those medicines is a very, very good argument indeed because we know xylitol to be non-cariogenic. Thus, we can give products of this type to children after they have brushed their teeth, provided they contain no other ingredients which may cause damage to the teeth. So these products are being marketed already and we have requests from the pharmaceutical industry to replace sucrose in other preparations as well, but syrups should certainly come first.

W. M. Nicol (Tate & Lyle Ltd, Reading University, UK): I was surprised to hear from Dr Doty that he has almost given up the fight against finding a cheaper method of producing xylitol, because I think it is a very interesting commodity which would be used very much more if the price is kept down. I would like to hear his views on how much the volume has to increase in order to make an official practical price. Is there really no other method of getting xylitol?

T. E. Doty (*Xyrofin Ltd, Baar, Switzerland*): I wish that I could give you a really concise answer. There are other methods of producing xylitol. Unfortunately, the other methods that are known today are more expensive than the one now being used. I did indicate that, because of the high investment involved in research leading to the production method used now, and in the plant necessary for the rather complicated process of production, the unit cost of the final product is high. It would be presumptuous of me to try to give you some kind of curve relating production increases to price decreases. We can say with certainty that as production does increase the price will come down. How far it comes down in the end is to some extent a function of how great the size of the production unit will be. I should say that we at Xyrofin are deeply involved with the problems of production technology. Mr Aminoff, the senior author of our paper, is involved full-time on this and increased production is imminent but to make any useful predictions on what the curve will look like in the long run will depend on the capacity. To give you a more hopeful view, let us consider the price of a sweetener related to level of sweetness as a measuring stick. The question of sweetness is a very relative thing, anyway, depending on concentration, temperature, pH and lots of other things. Assuming that we can find some standard of sweetness more or less acceptable to all people concerned, I think we can say that, instead of being a high multiple of the price of sucrose as it is at the present time on a pound-for-pound basis, we will find xylitol approaching a competitive level on the basis of sweetness with some of the other sucrose substitutes, namely mannitol and sorbitol. I hope that answers your question.

A. W. Holmes (*BFMIRA, Leatherhead, UK*): I really do not think that it does answer the question because it seems to me that the process of manufacture as you have described it, is relatively simple. After all, it is simple hydrolysis followed by hydrogenation, and that really is not complicated. There must be something else; is it the clean-up procedure that is expensive? After all, the oils and fats industry are regularly hydrogenating and producing quite a cheap product and hydrolysis is not expensive. Perhaps it is the pharmaceutical industry profit margin.

Doty: The production technology happens to have been developed by the Finnish Sugar Co. which is involved in the food industry. The Finnish Sugar Company's philosophy for its production of sucro-chemical specialities has always been that they would not be in the pharmaceutical market but that they would be food raw materials producers. We are realists enough to know that to be competitive in the food market you have to be competitive with the competing sweeteners that are there: we are very much aware of this. Now it did not answer the question for you and I am sorry that Mr Aminoff is not here to explain more fully. The scheme that we showed in the slide is essentially a simple straight-forward process. However, the sucro-chemical technology using ion-exchange and ion-exclusion chromatographic techniques for separating various sugars or other sugar-like substances from one another does involve a bit more than just simply hydrolysis and hydrogenation. I hope that answers your question a little bit more. There is a tremendous amount of patent literature available which is indicative of the complication of the process of getting a pure substance when you are starting out with a very heterogenous mixture.

Question: My other point really was a comment. I think that bacteria are more adaptable than man and the fact that man involves xylitol in his metabolism indicates that man can transform xylitol in some way. I suspect therefore that bacteria, given a chance, will find a way to do so. It worries me, therefore, when you say that 40 % marmalade for example can be stable bacteriologically because maybe the organisms have not had a chance yet and maybe the same point applies to the ability of xylitol to be metabolised in the oral cavity. Have any adaptation experiments been done in order to take account of that?

Voirol: Yes, I believe there have but I would like to ask for your patience until 14.30 this afternoon, when Professor Gehring will explain exactly what he did.

Question: According to your hand-out, the legal status of this material in the UK is that it is not actually listed anywhere, and therefore it is not currently permitted. Could you comment on any likely changes in that situation?

Doty: I think we would like to ask Mr Counsell to field that question if we may elect him, since he has been our co-ordinator for legal problems in the UK.

J. N. Counsell (*Roche Products Ltd, Dunstable, UK*)*:* It is rather difficult to make any sensible comment at this time because we have made a submission to the Ministry for its inclusion in a list. I have no idea into which list they propose to put it. I am only hoping that they will put it into one of them. As things stand at the moment, it is thought by some people at least to be a humectant although if you read the definition and look at the properties of humectants you find it rather hard to reconcile the two. It is not an artificial sweetener and, in fact, it fits into no category amongst those for food additives which are listed. Therefore, the question is, 'what is it?' This is, I think, the problem which the Ministry faces at the moment. But we certainly made our submission asking for it to be included in some list. We can only wait and see what happens. I am sorry it is not possible to give you a very definite answer. In other countries, it is I know, quite easy to do so.

Comment: The company I work for declares ingredients on labels and obviously we could not consider offering xylitol, even for consumer tests, unless we could so label it and so you can see our difficulty I think.

Counsell: Indeed I do.

Doty: If I may comment on this. If I understand correctly, the submission that has been made to the Ministry and the way that they requested the submission for the application for approval was, well, there were several possibilities. One would be just as a food substance which I believe the Ministry did not agree with. Even though it occurs naturally does not prove that it is a natural food substance as defined. I think, if I am not entirely mistaken, that the approach that has been taken is to make an application for approval of xylitol as an additive. It is not for us but for them to decide which category it would be in and I think the resolution of the problem would then be by adding it to the list of miscellaneous additives without really worrying about the question of whether it is a humectant or whatever.

A. E. Bender (Queen Elizabeth College, London University, UK): Would Mr McLachlan like to make a comment, I know he is an authority in this area.

T. McLachlan (98 *Boundary Road, London, UK*)*:* I think that the position of xylitol is rather interesting because if it is used as a sweetener it is not used as a humectant. The *Artificial Sweeteners in Food Regulations* state quite categorically that polyhydric alcohols are not artificial sweeteners. It then defines the polyhydric alcohol as an alcohol with three or more hydroxyl groups. Therefore, in my opinion if xylitol is used as a sweetening agent alone, it is permitted. But if it is used for any other purpose as a food additive, then I submit that it is illegal and this is one of the incomprehensible things about British law.

Question: This is not a technical question but I think many of us regard sucrose as a drug of addiction. I do not know whether you agree with this, and I am wondering whether your current studies on xylitol give any indication as to whether it is less addictive compared with sucrose. We have this problem with the small children. I was a little disappointed to find it had almost the same calorific value.

Doty: I think that this is a question that could be argued with a great deal of fervour from either point of view. I have heard arguments on both sides, and I think I agree with both sides. Maybe it boils down to a question of whether the sweet taste is inherited or developed. The research that has been done on sweetness *per se* has shown that by injecting various sweet substances into amniotic fluid, for example, will increase the sucking reaction of the foetus showing that there is a positive response to the sweet taste. I think that Mother Nature had a reason for arranging that because mothers' milk is actually quite sweet. Perhaps it is the way that nature has provided for making sure that the newborn infant or animal, when he found the source of sweetness, would stick to it and take his fill while becoming addicted to sweetness. I think that Mr Kracher also gave us a good example from the time when man was a forager that sweetness was good but if it was sour then man tended to be a little bit more suspicious of it. I think the question of whether or not one becomes addicted to sweetness really is not pertinent to this discussion. On the other hand, I do think that the question of whether one sweetener or another is preferable for certain uses is pertinent. The evidence that we will hear today and which has already been published will justify the statement that each of the sweeteners has its own law, whether it is a synthetic, artificial, calorie-free, a relatively less sweet starch hydrolysate, sucrose, sorbitol, mannitol or xylitol I think that a good justification can be made for each of these sweeteners. They do have definite rôles to play and they are not exclusive of each other.

Bender: It would appear, in fact, that the addiction, if it is an addiction, is for sweetness and not for sucrose rather than something else. So if one is a sweet addict then this could be an addiction to any sweetener.

Doty: I think so.

M. C. Kane (Spillers Research & Technical Centre, Cambridge, UK): The use of xylitol would seem to be limited by the cariogenicity of other ingredients in foodstuffs. Did the dietary studies indicate any relative cariogenicity of the main classes of foodstuffs, i.e. relative to sucrose at one end of the scale and xylitol at the other?

Doty: Some of the other papers that we will hear later today will deal more fully with relative cariogenicity. Very roughly, the Turku dietary study showed that with complete substitution of sucrose with either fructose or xylitol some kind of scale could be established. Taking sucrose as highly cariogenic, fructose was approximately 30% less causative of new caries and xylitol was non-causative of new caries. That indicates the scale and that is just with regard to sugars or sweeteners. As to other foodstuffs, I doubt whether there is much evidence from the dietary studies in Turku to indicate, for example, whether bread was more cariogenic than breakfast cereal.

Question: Firstly, if you are making representations to the Ministry of Food, I do hope that someone is also looking after the British Diabetic Association, which from our experience with fructose could take longer. It would also be important. Secondly, is there a more homely synonym that we can use? There is 'fruit sugar' for fructose but I do not think much of 'wood sugar' for xylitol. If it is to be a main food ingredient in some foods, a name like xylitol will be rather difficult for informing the public. It is therefore against the chances of marketing it successfully.

Bender: The concentration is rather high in the aubergine. How about 'egg sugar?'

Doty: In Finland and, I think, in Sweden and Norway as well, xylitol is known in the vernacular as 'birch sugar' because our raw material at present is birch wood. It is misleading because it is not a sugar and it is not a component, as such, of birch. However, there obviously has been a lot of publicity about the development of the production and so on and thus birch sugar is just a folk term for it. If, for example, cotton-seed hulls or oat hulls or nut-shells were used as the raw material would, say, nut-shell sugar be much preferable to that? On the other hand, xylitol as a name, again because of the publicity of the development of the production technology and also of the Turku studies, is itself a very acceptable and marketable term in Finland. It is a matter of information. When the public hears and knows and is informed of what xylitol is then it does have some kind of a ring itself. One of the most common marketing techniques with xylitol-containing products in Finland, and I think this could be true in the rest of Scandinavia today, is to have xylitol appear on the pack as part of the name of the product. It is a generic name so it is not registered, anybody can use it. Now to extrapolate from that to other countries I would say that it requires a great deal of good valid information so that when people hear the term something will click in their minds to tell them that this is a favourable or preferable alternative.

S. Leach (Dept. Dental Sciences, Liverpool University, UK): Several of my dental colleagues who, I think, have an emotional rather than a rational suspicion against xylitol at this time, keep reminding me of the dangers inherent in there being trace amounts of xylose still present in the xylitol. Could you tell us something about that, please?

E. Vanninen (*Xyrofin Ltd, Baar, Switzerland*): I think I would like to touch on two subjects, actually. The question of production costs of xylitol: being partly in charge of the production people in Finland, this question, of course, is close to my heart. I can only try to convince you that we do all that we can to reduce production costs of xylitol. Unfortunately, it is just not a question of simple hydrolysis and hydrogenation. As a raw material, we use birch wood and we can easily get a yield of about 20 % or less of the dry substance in wood which means that we produce a lot of by-products as well. Because of the method of hydrolysis we use, the by-product (mainly cellulose) cannot be used as normal cellulose. It must be mixed with other products and processed further to be commercially useful. In the hydrolysis process, we unfortunately produce a lot of material which cannot just be abandoned unless they are purified, otherwise we would pollute our environment. I am talking about substances like acetic acid, concentrated salt solutions which are a part of the hydrolysis and the hydrolysate itself is a mixture of sugars, it is not pure xylose. It consists of glucose, galactose, arabinose, even some fructose and so on. It must be purified, it must be desalted and it must be purified again to obtain a solution that can be hydrogenated which, however, still contains some impurities. After hydrogenation we get a mixture of polyalcohols which still contain some reducing sugars. Among them possibly even xylose. I would like to revert to that question later. The product must be once more purified. We use mainly ion-exclusion ion-exchange technology for this purpose, and I have to tell you that the investment that has been made is really high and the process itself requires very much energy which, in our days, is very expensive, unfortunately. I agree with Dr Doty when he says that production costs will be reduced in the future. I am a bit pessimistic about whether they can be reduced from the present price very much. I hope we can keep the price at its present level which means of course that it agrees in comparison with other raw materials. One of the requirements of any food ingredient today is of course a maximum of purity and I have seen, for instance, xylitol that is made and produced in China which could not be approved by any authorities in the Western world. It would be really cheap to produce this kind of material and of course we could do it at a pretty low price but it could not be used as a food ingredient. One of the impurities that theoretically could be available in xylitol is xylose, the sugar from which xylitol is actually derived. The hydrogenation process however prefers xylose before glucose or other sugars which means that xylose will be totally hydrogenated in the process. On the other hand, the material is purified afterwards so that the total amount of reducing sugars is below 0·01 %. Theoretically, a part of this could be in the form of xylose. Practically, however, there are no detectable amounts of xylose present in our xylitol.

CHAPTER 4

Biochemistry of Xylitol

K. H. BÄSSLER
(*Johannes Gutenberg University, Mainz, W. Germany*)

While xylitol was first prepared by the reduction of D-xylose as early as 1891 by E. Fischer and also by Bertrand, more than half a century had passed before we learned that it is a substance of biological importance. Today we know that it is of widespread occurrence in human foods such as fruits, vegetables, cereals and mushrooms. In 1955, Touster detected xylitol as a normal intermediate in human metabolism in the course of investigations on essential pentosuria, an inborn error of metabolism (Touster, 1960; Touster, Reynolds and Hutcheson, 1956). These investigations led to the discovery of a glucuronic acid–xylulose cycle (Fig. 1) which presumably has the function of recycling glucuronic acid, not used for synthetic reactions or biotransformations, into glucose metabolism. In this metabolic pathway xylitol is the intermediate in the conversion of L-xylulose to D-xylulose taking place in liver mitochondria.

In essential pentosuria the dehydrogenase which converts L-xylulose to xylitol is missing, so that L-xylulose accumulates and is excreted in the urine. The glucuronic acid–xylulose cycle is closely connected with the pentose phosphate shunt as shown in Fig. 2. It enters this pathway by phosphorylation of D-xylulose to D-xylulose-5-phosphate.

At this stage our group in Mainz tested the metabolic turnover

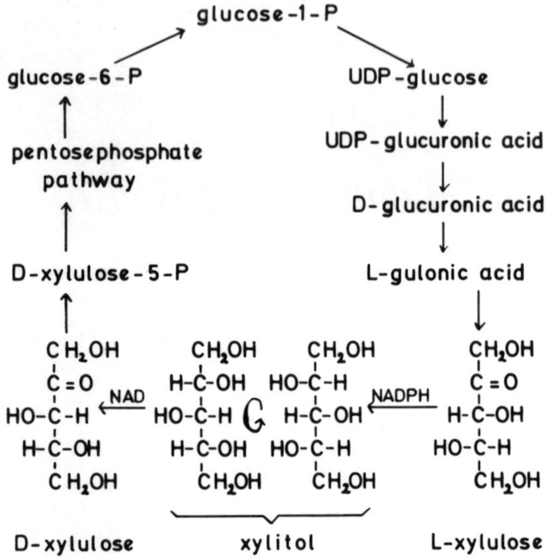
glucose-1-P
glucose-6-P
UDP-glucose
pentosephosphate pathway
UDP-glucuronic acid
D-glucuronic acid
D-xylulose-5-P
L-gulonic acid
NAD
NADPH
D-xylulose
xylitol
L-xylulose

FIG. 1.

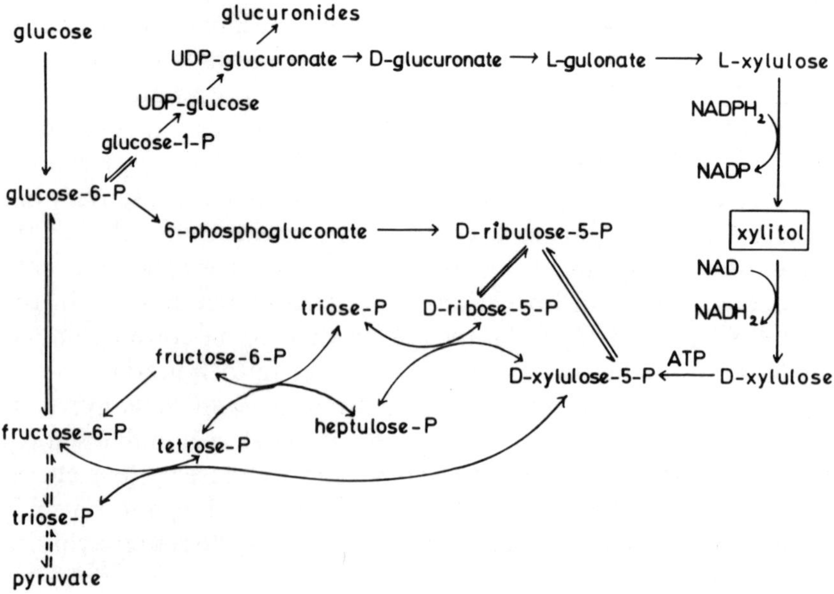
glucose
glucuronides
UDP-glucuronate
D-glucuronate
L-gulonate
L-xylulose
UDP-glucose
glucose-1-P
NADPH₂
NADP
glucose-6-P
6-phosphogluconate
D-ribulose-5-P
xylitol
NAD
NADH₂
triose-P
D-ribose-5-P
fructose-6-P
ATP
D-xylulose-5-P
D-xylulose
heptulose-P
fructose-6-P
tetrose-P
triose-P
pyruvate

FIG. 2.

capacity for exogenously administered xylitol in animals and human beings. It was found that an adult man of 70 kg can metabolise some 500 to 600 g of xylitol per day when it is infused intravenously. This was in sharp contrast to the capacity of Touster's cycle of 5 to 15 g per day. Therefore, another pathway had to exist for exogenous xylitol and we were able to show that administered xylitol is

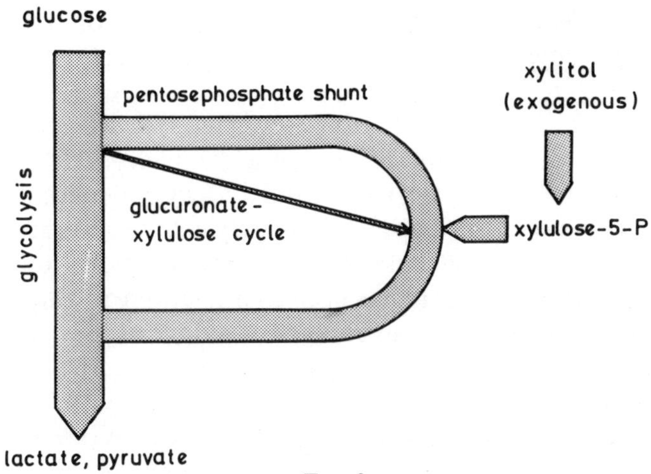

FIG. 3.

dehydrogenated very efficiently to D-xylulose by the cytoplasmatic enzyme L-iditol dehydrogenase which also reacts with sorbitol and some other polyols (Bässler, Stein and Belzer, 1966). D-Xylulose then enters the pentose phosphate shunt by phosphorylation to D-xylulose-5-phosphate. This is shown schematically in Fig. 3 where the width of the arrows symbolises the turnover capacity of the pathways.

It was thus evident that xylitol metabolism is of an order of magnitude that is energetically important. The sweet taste of the substance and its favourable technological properties moreover suggested its use as a substitute for sucrose or glucose. If we consider such a rôle for xylitol, we have to know more about its metabolic behaviour, its relevance for energy production, the distribution of xylitol metabolising enzymes in the organism and the rate of absorption from the gut. A large body of knowledge in this field has accumulated during the past 20 years. The main results can be briefly summarised as follows.

If xylitol is administered either orally or by the intravenous route, about 85 % is metabolised by the liver, 10 % by the kidneys and the small remainder is used up by blood cells, the adrenal cortex and a number of other tissues of minor importance. This was shown by enzyme studies (Bässler, Stein and Belzer, 1966), infusion experiments (Müller *et al.* 1967), and recently confirmed and extended by studies with radioactive-labelled xylitol (Quadflieg and Brand, 1976).

In the liver, xylitol is partly converted to glucose, partly metabolised to pyruvate and lactate, or is completely oxidised. The portion converted to glucose depends on the metabolic situation, for instance on the need for glucose. It may vary between 20 and 80 % of the total turnover. This glucose production proceeds smoothly with retardation (xylitol therefore has been characterised as 'glucose with delay') which is one of the advantages of xylitol as a sweetener of foods for diabetic patients. In this way it acts like starch compared with glucose or sucrose.

It is certainly not completely correct to characterise xylitol as 'independent of insulin'. The slow and smooth glucose production, however, causes only a very low stimulation of insulin secretion. This is, in addition, a dietetic advantage since high insulin levels stimulate the appetite and the undesired conversion of carbohydrates into fat.

In situations such as starvation, diabetes or stress, xylitol lowers free fatty acid levels, acts antiketogenically and improves the peripheral glucose utilisation without additional need of insulin (Haydon, 1961; Bässler and Dreiss, 1963; Toussaint, Roggenkamp and Bässler, 1967; Humke and Wiessmann, 1971; Grabner *et al.*, 1971; Hammerl *et al.*, 1971; Förster and Zagel, 1974; Förster, Haslbeck and Mehnert, 1974).

A balance of energy production during xylitol metabolism is presented in Fig. 4. For each kilocalorie of its physical combustion value, xylitol furnishes 0·056 moles of ATP, exactly as much as glucose. Therefore, xylitol should not be mistaken as a non-caloric sweetener. It supplies as much energy per gram as sucrose and this should be taken into account when dietetic regimens are designed.

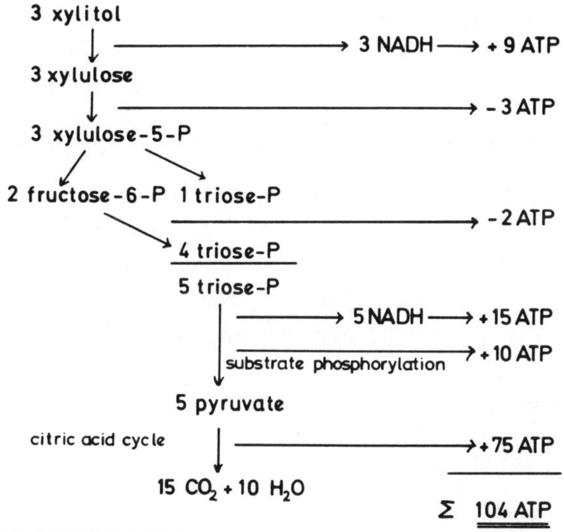

	xylitol	glucose
mol ATP/mol	34·67	38
mol ATP/g	0·228	0·211
mol ATP/kcal combustion value	0·056	0·056

FIG. 4.

The absorption of xylitol from the intestine is a rather slow process occurring by passive diffusion. This limits the amount applicable by the oral route since excess doses may cause osmotic diarrhoea. Although adaptation to amounts up to 200 g per day is possible (Dubach, Feiner and Forgó, 1969), usually not more than 50–70 g daily, spread evenly throughout the day, should be given. Due to this natural low limit, all potential side effects of xylitol can be completely disregarded as long as it is consumed orally. The situation is different if xylitol is infused intravenously; there is no natural barrier, like absorption, in this route and the organism can be flooded with xylitol if the rate of infusion exceeds the turnover rate. This may cause undesired effects like a shift in the cytoplasmatic redox state, decrease of liver adenine nucleotides with trapping of phosphate in metabolic intermediates (Woods and Krebs, 1973), increased uric acid production and mineral losses due to enhanced diuresis. Therefore, we have to set limits. As a result of very intensive studies in healthy and sick subjects, an infusion rate of 0·25 g xylitol per kg in an hour has been recommended as a safe limit (Berg, Matzkies and Bickel, 1974; Bässler, 1976). This value has also been adopted by the German Drug Commission. I shall, however, not dwell further on parenteral problems since the subject of this symposium clearly is centered on the oral use of xylitol.

In summary, from our knowledge of xylitol metabolism and from experiments in animals and humans, we can conclude that xylitol is metabolised mainly by the liver where it is partly oxidised to yield energy and partly converted to glucose. Since the conversion to glucose proceeds smoothly with no abrupt effects on the blood glucose level and insulin secretion, xylitol instead of sucrose in the diet of diabetics makes metabolic control easier. In starvation, diabetes and stress situations, xylitol has a normalising effect on fat metabolism, lowering free fatty acid levels and ketone production, and in this way improves glucose utilisation without need of additional insulin.

REFERENCES

Bässler, K. H. (1976) Quantitative aspects of the metabolism of glucose, fructose, sorbitol and xylitol. In: *Monosaccharides and Polyalcohols in Nutrition, Therapy and Dietetics*, eds. G. Ritzel & G. Brubacher, *Int. Z. Vit.-u. Ernährungsforschg.*, Beiheft **15,** pp. 22–30. Bern–Stuttgart–Wien: Hans Huber.

Bässler, K. H. & Dreiss, G. (1963) Antiketogene Wirkung von Xylit bei alloxandiabetischen Ratten. *Klin. Wschr.*, **41,** 593–5.

Bässler, K. H., Stein, G. & Belzer, W. (1966) Xylitstoffweschsel und Xylitresorption. Stoffwechseladaptation als Urasche für Resorptionsbeschleunigung. *Biochem. Z.*, **346,** 171–85.

Berg, G., Matzkies, F. & Bickel, H. (1974) Dosierunsgrenzen bei der Infusion von Glucose, Sorbit, Fructose, Xylit und deren Mischungen. *Dtsch. med. Wschr.*, **99,** 633–8.

Dubach, U. C., Feiner, E. & Forgó, I. (1969) Orale Verträglichkeit von Xylit bei stoffwechselgesunden Probanden. *Schweiz. med. Wschr.*, **99,** 190–4.

Förster, H. & Zagel, D. (1974) Stoffwechseluntersuchungen während und im Anschluss an Dauerinfusionen von Glucose und von Zuckeraustauschstoffen. *Dtsch. med. Wschr.*, **99,** 1300–4.

Förster, H., Haslbeck, M. & Mehnert, H. (1974) Die Bedeutung der Kohlenhydrate in der parenteralen Ernährung. *Infusionstherapie*, **1,** 2–40.

Grabner, W., Berg, G., Bergner, D., Matzkies, F. & Maeder, H.-U. (1971) Über den Einfluss des Xylits auf den Fettstoffwechsel von Diabetikern. *Med. Ernähr.*, **12,** 126–8.

Hammerl, H., Kränzl, Ch., Nebosis, G., Pichler, D. & Studlar, M. (1971) Therapeutische Aspekte der Stoffwechselwirkungen von Xylit bei Diabetikern. *Wiener Z. Inn. Med.*, **52,** 70–4.

Haydon, R. K. (1961) The antiketogenic effects of polyhydric alcohols in rat-liver slices. *Biochim. Biophys. Acta*, **46,** 598–9.

Humke, W. & Wiessmann, K.-J. (1971) Die peripartale Ketonämie und ihre Beeinflussung durch Xylit. *Arch. Gynäkol.*, **210,** 97–100.

Müller, F., Strack, E., Kuhfahl, E. & Dettmer, D. (1967) Det Stoffwechsel von Xylit bei normalen und alloxandiabetischen Kaninchen. *Z. ges. exp. Med.*, **142,** 338–50.

Quadflieg, K.-H. & Brand, K. (1976) Comparison of xylitol and glucose metabolism in nonhepatic rat tissues. *Z. Ernährungswiss.*, **15,** 345–54.

Toussaint, W., Roggenkamp, K. & Bässler, K. H. (1967) Behandlung det Ketonämie im Kindesalter mit Xylit. *Z. Kinderheilkunde.*, **98,** 146–54.

Touster, O. (1960) Essential pentosuria and the glucuronate–xylulose pathway. *Fed. Proc.*, **19,** 977–83.

Touster, O., Reynolds, V. H. & Hutcheson, R. M. (1956) The reduction of L-xylulose to xylitol by guinea pig liver mitochondria. *J. Biol. Chem.*, **221,** 697–709.

Woods, H. F. & Krebs, H. A. (1973) Xylitol metabolism in the isolated perfused rat liver. *Biochem. J.*, **134,** 437–43.

CHAPTER 5

Tolerance in the Human. Adults and Children

H. FÖRSTER

(*Klinikum der Johann Wolfgang Goethe University, Frankfurt-am-Main, W. Germany*)

The metabolism of the polyalcohol xylitol has been studied in detail in recent years. The results of these investigations led initially to its use as an energy source in parenteral nutrition. In addition, research was undertaken into its potential for enteral use. This showed that xylitol might be indicated not only for diabetes mellitus but also for the prevention of caries. One significant question which attends the oral use of polyols, including sorbitol and mannitol, is the—as yet little researched—intestinal absorption of these substances. Similarly, there are very different opinions concerning the oral tolerance of these substances.

The pentitol xylitol is a highly hydrophilic and thus lipophobic substance which, because of its molecular weight of 152, can certainly not be absorbed very quickly through hydrophilic pores from the intestinal lumen, unless some form of transport process is present. The exceptionally good intestinal absorption of certain sugars despite higher molecular weights can be explained primarily by the presence of transport processes. One should, therefore, first discuss the various possibilities of absorption of hydrophilic substances from the intestinal lumen.

INTESTINAL ABSORPTION OF HYDROPHILIC SUBSTANCES

There is, of course, only basic knowledge of the structure of biological membranes, and numerous questions remain unanswered (Packer, 1974; Wright, 1974). Even the question of the existence of hydrophilic pores has not yet been answered definitely. The orthodox view is that the structures of biological membranes consist

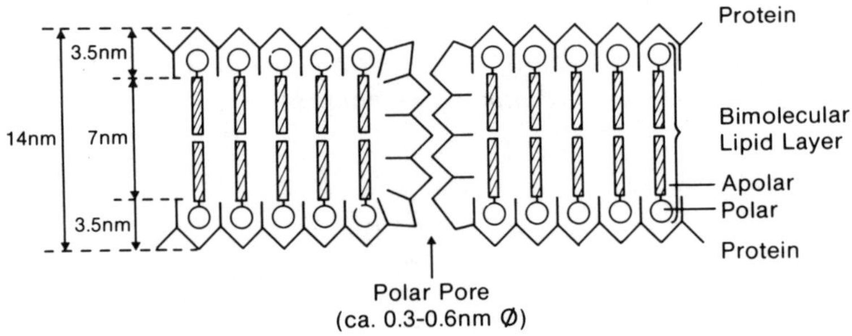

FIG. 1. Schematic representation of the cell-membrane structure.

of hydrophobic lipid double layers containing hydrophilic pores (Fig. 1). The thickness of the membrane lies in the region of 10 nm, whereas the diameter of hydrophilic pores can clearly only range between 0·3–0·6 nm, i.e. <1 nm. Consequently, we must consider tunnel-like structures rather than pores, as their length exceeds their diameter by a factor of 10. For structures of this type, certain characteristics apply (for example, single-file diffusion) but on the other hand, the laws of diffusion must be applied with certain restrictions (Wright, 1974; Packer, 1974). The difference in concentration provides the motive power for free diffusion of a substance. The net transfer of substance is, over a wide range, directly proportional to the difference in concentration, in so far as there are no larger differences in the osmotic pressure on either side of the biological membrane. A significant feature, therefore, of free diffusion is that as no substrate saturation can be established, there is no limiting value for the substrate transfer. An increase in the concentration levels of the substance leads to an increase in transfer through the membrane. It is only when the concentrations which can be tolerated by the system being studied are exceeded that this simple relationship is disturbed.

The second biological possibility for the explanation of the passage of hydrophilic substances through lipid membranes is described as 'facilitated diffusion'. The starting point of this is that a hitherto hypothetical carrier molecule ('carrier') enters into a reversible association with the corresponding substance (Wilson, 1962; Csaky, 1975). The hydrophilic properties of the substance which is to be transported are thereby masked and its passage through the hydrophobic membrane is made possible. As the number of carrier molecules within a membrane is limited, 'substrate saturation' can be established, parallel to enzyme kinetics. No energy-consuming uphill transport can be performed by this 'facilitated transport'. The direction of transport is determined solely by the direction of the concentration gradient; and the direction of transport is, therefore, reversible. In this, circumstances are similar to those of free diffusion. When substrate saturation is very high for the appropriate carrier system, one can no longer distinguish simply between free diffusion and facilitated diffusion. In biological systems the usable concentration is limited by the osmotic pressure of the solutions. Concentrations of substance which are well above the physiological osmotic pressure of approximately 300 mosmole/l lead to the destruction of the membrane systems. The absorption of fructose probably takes place by means of facilitated diffusion. As far as the intestinal absorption of fructose is concerned, substrate saturation was established—within limits (Guy and Deren, 1971; Förster, unpublished). Moreover, the absorption capacity for fructose is significantly higher than for other substances with a similar molecular weight, as earlier experiments have shown (Cori, 1925).

In the case of active transport as the most highly developed biological system for passing a substance through membranes, the direction of transport is not determined by the concentration level of the appropriate substance, but by the molecular arrangement of the transport system within the membrane (Wilson, 1962; Förster and Menzel, 1972; Csaky, 1975). As in the case of facilitated diffusion, passage through the membrane takes place by way of a reversible combination with a possibly mobile carrier (Fig. 2).

The most significant difference as compared with facilitated diffusion is the possibility of transport against an electrochemical gradient, or in simpler terms, the possibility of transport in an uphill direction. Uphill transport, then, takes place by using energy

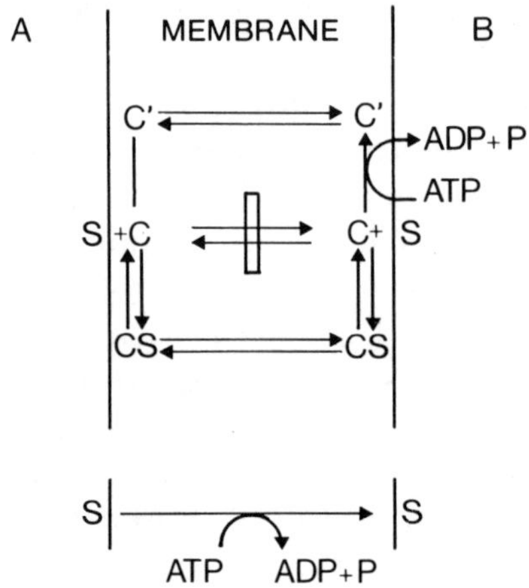

FIG. 2. Scheme of the active transport of the asymmetric carrier. The substance, S, is transported across a hydrophilic membrane from A to B by means of the carrier, C, which is activated, C′, by ATP.

because the movement in transport, according to the theoretical model, is connected with the splitting of a high-energy compound. This model allows, in principle, no reversal of transport direction. It is in this way that the active absorption of glucose and galactose from the intestinal lumen takes place (Wilson, 1962; Förster and Menzel, 1972).

For xylitol—and the same applies to sorbitol—no indication of the presence of transport mechanisms has been established to date. The absorption of both of these polyols would, therefore, take place exclusively by means of free diffusion or by way of transport systems, as yet undiscovered, with a low affinity to the polyols. It has already been mentioned that the pore size is a limiting factor for transport processes of this type. There is relatively little difference in the molecular weights of sorbitol and xylitol—that of sorbitol is 182, and that of xylitol only 152. This relatively slight difference of less than 20% must be of significance in the limits within which free diffusion operates. The molecular structure as a whole may well be of additional importance for the possibility of diffusion. Xylitol has a completely symmetrical structure. On the other hand, sorbitol has certain points of asymmetry—the hydroxyl groups at C-4 and C-5 are both arranged in a D-configuration. Additionally, there is the factor that relatively stretched molecules are poorly suited to processes of permeation.

In the case of glucose (and the same would apply to galactose), free diffusion is comparatively pronounced when correspondingly high concentrations are present in the intestinal lumen (Förster and Menzel, 1972). Glucose has a molecular weight of 180 (close to that of sorbitol), yet it occurs almost completely as a ring structure with a molecular weight of 162.

As far as the sugars are concerned, there is the possibility of almost completely blocking active transport by means of a specific inhibitor, phlorizin. The remaining absorption of glucose then occurs by way of free diffusion (Förster and Menzel, 1972). It reaches relatively large proportions in animal experiments (Fig. 3). The part played by free diffusion in absorption is dependent on the concentration of the sugar in the intestinal lumen. Where lower concentrations of glucose are involved, glucose absorption occurs almost exclusively by way of active transport. When sugar concentrations are correspondingly higher, the part played by diffusion may, under given physiological conditions, rise to around 50% of the total absorption of substance (Förster and Menzel, 1972). In man, corresponding experiments can only be carried out with difficulty. Nevertheless, under extreme conditions, as perhaps in a high-dosage oral glucose tolerance test, conditions can be achieved even in man, in which the absorption of glucose from the intestinal lumen by means of diffusion can reach notable proportions. As compared with the polyols, the very favourable molecular structure of glucose as β-pyranose (where all the hydroxyl groups are arranged in an equatorial position) makes this behaviour in free diffusion possible, which cannot be explained simply on the

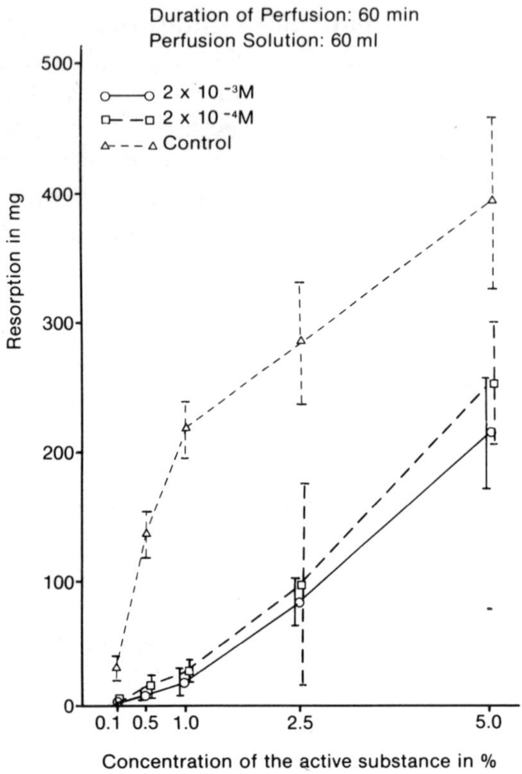

FIG. 3. The influence of phlorizin on glucose resorption.

basis of the molecular weight. Earlier experiments of our own, using not very specific methods, led to the conclusion that intestinal absorption of xylitol amounted to only about 35–60 % of the free diffusion of glucose (Fig. 4), although the molecular weight of glucose is 180 as compared with 152 for xylitol (Förster and Menzel, 1972). For sorbitol with a molecular weight of 182, one could calculate for its intestinal absorption from the intestinal lumen, about 50–70 % of the absorption of xylitol. However, there is, of course, still the uncertainty, which applies as much to sorbitol, in the non-specific determination of polyols in the experiments done at the time (Förster, unpublished). The data for the absorption of other, not transported, sugars and polyols from the intestinal lumen by

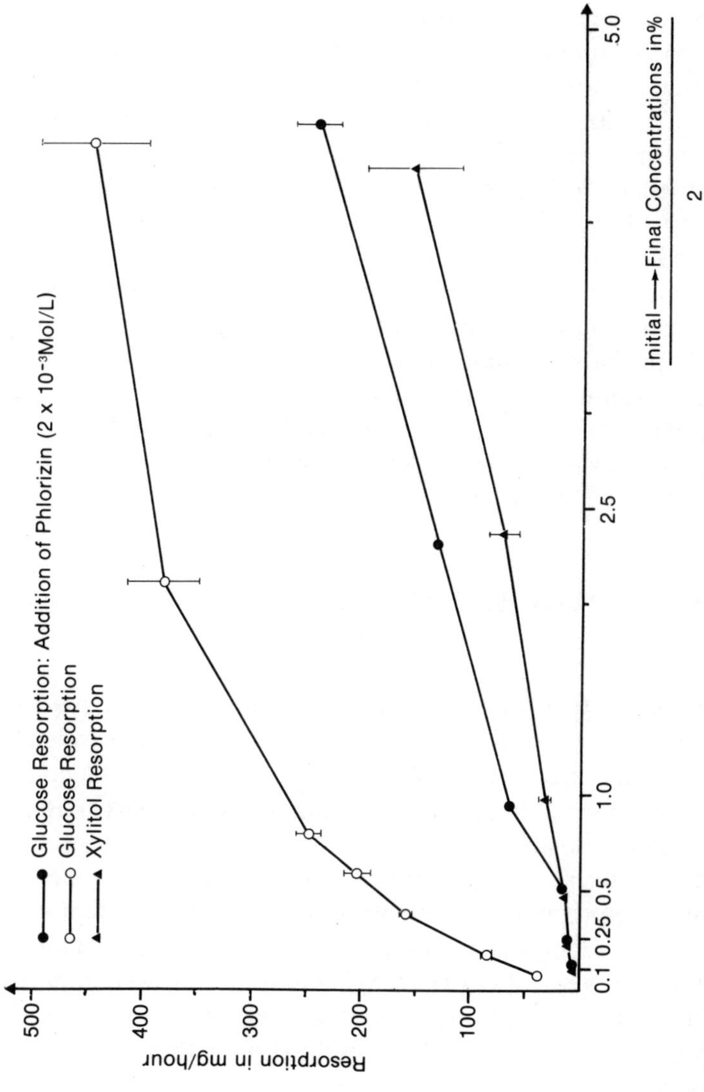

FIG. 4. Resorption of glucose with or without addition of phlorizin and xylitol from the perfused small intestine of the rat. Addition of 0·15 mol/l NaCl to the perfusion medium.

means of free diffusion conflict greatly. On the basis of these data, both of the hexoses, mannose and sorbose, should hardly be absorbed in man (Holdsworth and Dawson, 1964). But earlier experiments resulted in a pattern of intestinal absorption of sorbose in experimental animals, which could be regarded as typical of free diffusion (Verzar, 1935). Similarly, in experiments on chickens, it was established that mannose and sorbose are absorbed to a certain extent (Bogner, 1961; Bogner & Haines, 1961). These contradictions are probably brought about by the fact that these sugars can only be determined by non-specific methods.

In summarising, it should be emphasised that intestinal absorption by means of free diffusion in the case of xylitol and sorbitol is probably significantly below the level of free diffusion of glucose, on account of molecular structure. In addition, there is the active transport of glucose in absorption from the intestinal lumen, which achieves similarly high levels, dependent, of course, on the concentration of the sugar.

COMPARISON OF THE ABSORPTION OF SUGARS AND POLYOLS

As early as the turn of the century, the absorption of sugars (or monosaccharides) was divided into various groups, independently of the fundamental mechanism (Wilson, 1962). Glucose and galactose were regarded as rapidly absorbable sugars, fructose was assigned to an intermediate position, and the pentoses were regarded as poorly absorbable. These early studies yielded, thereby, the clear conclusion even then, that absorption from the intestinal lumen was almost independent of the molecular weight of the sugar. The detailed experiments of Cori (1925) led to quantitative comparisons of the different absorption rates. The technique on which these experiments were based relied on application with a stomach tube. The animals were subsequently killed at predetermined intervals, and the sugar residues remaining in the intestine were determined. A short time later comparable experiments were carried out by Verzar and Laszt (1935), employing direct application of sugars to a tied loop of intestine. The resulting relationships were similar, although emptying of the stomach, which has a regulating function in the case of oral application, was of course excluded in the case of direct application to the tied loop of the intestine. The experiments were repeated numerous times, but the relationships were always comparable to a high degree (Wilson, 1962; Förster, 1962). Later,

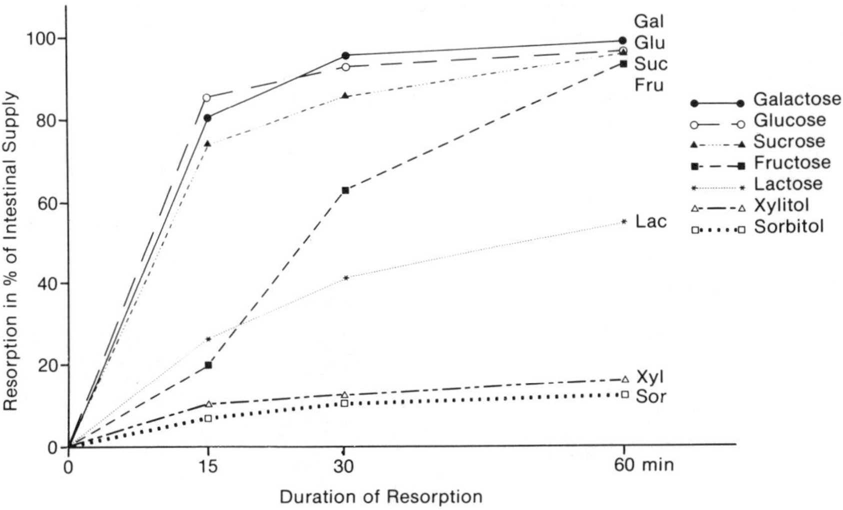

FIG. 5. The percentage resorption of 10 % solutions (1·3 ml) of galactose, glucose, fructose, sucrose, lactose, xylitol and sorbitol by rats.

using a modification of the Cori technique, corresponding values were also determined for the polyols, to compare with the Cori technique (Förster, 1962; Mehnert and Förster, 1968) (Fig. 5).

It proved to be extraordinarily difficult to find a suitable method for experimentation on man. Early studies using an intestinal double balloon tube resulted, in comparison, in considerably higher absorption rates for the small intestine (Mehnert *et al.*, 1959) than had been established in animal experiments. Subsequent experiments using a non-absorbable control substance produced significantly lower values (Fig. 6), although the results were still distinctly higher than those in the animal experiments (Mehnert *et*

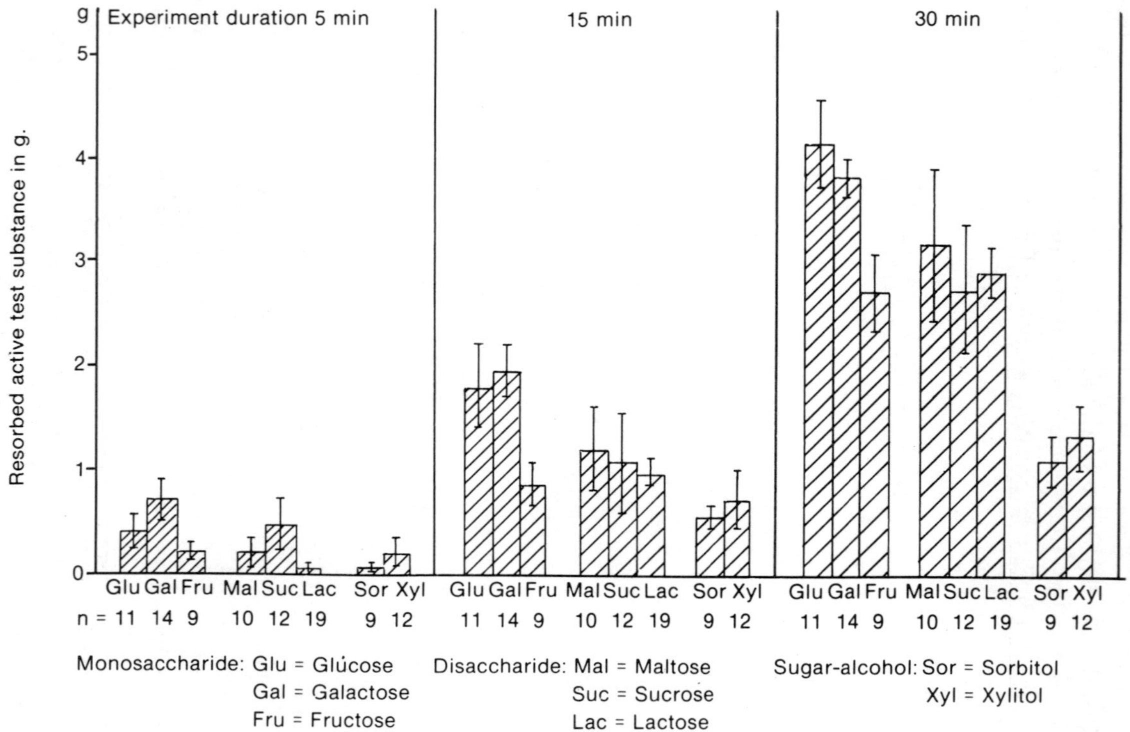

FIG. 6. Resorption of various sugars and sugar alcohols from a section of the human small intestine isolated by a double ball probang; 5 g of each active test substance. (Initial concentration 5 %.)

al., 1959; Mehnert and Förster, 1968; Liebau, 1964). These differences might be due to experimental shortcomings.

SIGNIFICANCE OF STOMACH EMPTYING FOR ABSORPTION

The different absorption mechanism of glucose (and galactose) on the one hand, of the polyols on the other, and between them of fructose, was shown by animal experiments using the stomach tube (Fig. 7). Glucose (and likewise galactose) was almost completely absorbed in the upper small intestine. No glucose at all reached the lower two-thirds of the small intestine following application with the

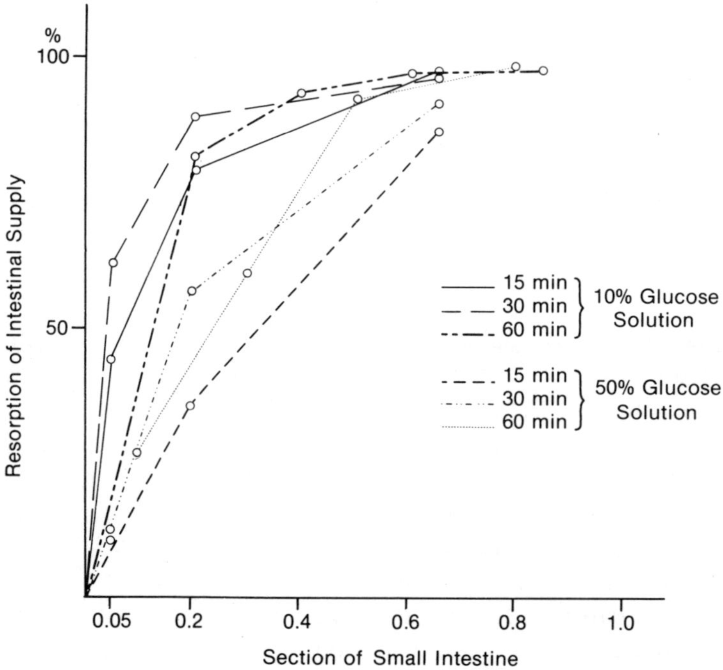

Fig. 7. Dependence of resorption of 10% and 50% glucose solutions on the duration of resorption and localisation in various sections of the small intestine in the rat (1·0 = total length of small intestine).

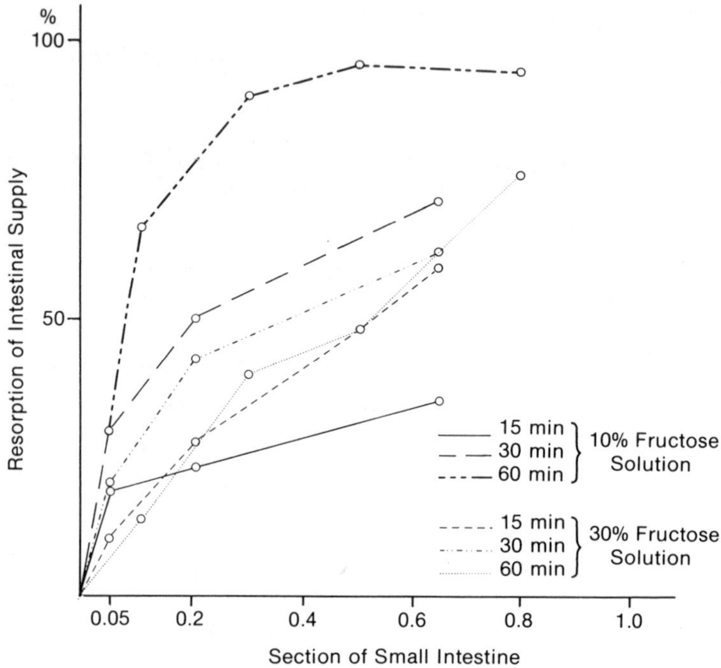

FIG. 8. Dependence of resorption of 10 % and 30 % fructose solutions on duration of resorption and localisation of resorption in various sections of the small intestine in the rat (1·0 = total length of small intestine).

stomach tube. This behaviour is the result of 'active absorption' of glucose by means of a 'carrier system' and processes working in close conjunction with the physiological mechanism of gastric emptying (Mehnert and Förster, 1968). If fructose is given to experimental animals with a stomach tube, there are still distinct quantities of fructose to be found in the lower regions of the small intestine (Fig. 8). When polyols are used, which leave the intestinal lumen exclusively by free diffusion, still greater quantities of these substances can be determined in the lower regions of the intestine after application with a stomach tube. When pure solutions are used, they are adjusted in the intestinal lumen to the osmotic conditions of the organism, i.e. solutions of about 300 mosmole/l result. According to the original concentration (osmolarity) of the solution,

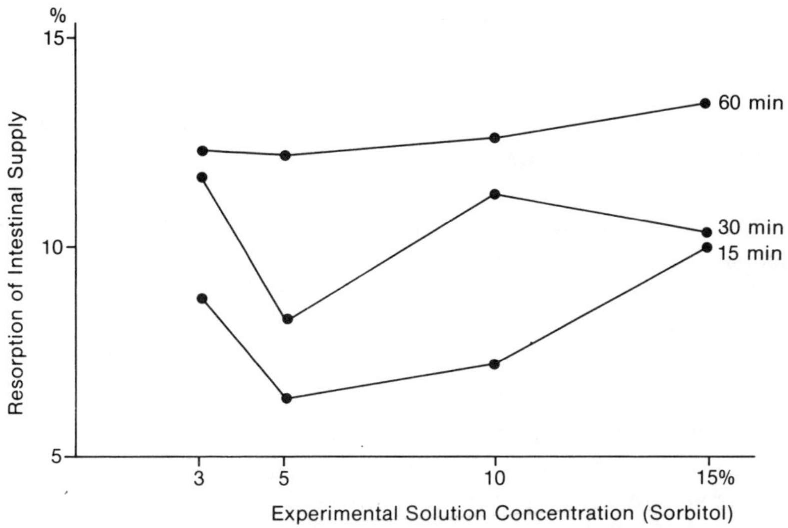

FIG. 9. Dependence of the percentage of sorbitol resorbed on the concentration of experimental solutions and the duration of the test period in rats.

either absorption processes or secretion processes take place. When a 10 % solution of sorbitol or xylitol is given with a stomach tube, the solution is first diluted in the intestinal lumen to about 3–4 % and then the polyol is absorbed together with water. The concentration of polyol is still in the region of 3–4 % even in the lower small intestine, in such circumstances (Fig. 9). As compared with Figs. 7 and 8, large amounts of the polyol are found in the lower small intestine in spite of the initially lower concentration of the polyol compared with the sugars.

The animal experiments with the stomach tube led to the conclusion that polyols are absorbed relatively slowly in rats. In contrast with the actively absorbed sugars, glucose and glactose, or with fructose which is absorbed by facilitated diffusion from the intestinal lumen, the possibility exists that at a higher oral dosage, polyols not absorbed in the small intestine can reach the large intestine. This has subsequently been proved in the case of sorbitol in man (Förster, unpublished). Similarly this seems to apply to xylitol, following the experiments carried out by Asano (Asano, Levitt and Goetz, 1972).

When concentrated solutions are used (and this applies for meals too) intestinal absorption is limited by stomach emptying. The presence of greater amounts of substances in the small intestine leads to inhibition of gastric emptying. All in all, filling the stomach represents a stimulation to stomach emptying, whereas filling the small intestine inhibits gastric emptying. In the case of rapidly absorbed sugars (glucose or galactose) the interplay of the different mechanisms is so good that absorption from the intestine is limited solely by the regulation of gastric emptying. The maximum absorption capacity cannot be fully used as long as the regulatory processes are intact. It is therefore impossible, even with quite large oral dosages of glucose, to induce diarrhoea. In our own experiments, 400 g of glucose were tolerated by human volunteers without difficulty (Förster, Meyer and Ziege, 1972).

Rats voluntarily drink up to 10 ml of a 20 % solution of glucose within 30 min (Förster, unpublished). As the animals' body weight was 200 g, the consumption was 10 g/kg body weight. Scaled up to human proportions, this would represent 600–700 g of glucose, and the quantities are, therefore, quite comparable. In experimental animals, glucose can still be found in the stomach as long as 3–4 h after ingestion. In human subjects, absorption of 300 or 400 g of glucose is only absolutely completed after 5–8 h (Förster, Boecker and Ziege, 1972; Haslbeck, 1976).

It is, therefore, scarcely possible to induce diarrhoea in a healthy human subject using the rapidly absorbable sugars, glucose or galactose. In the case of fructose, although the threshold value is relatively high, one must, of course, reckon with an osmotic diarrhoea if the value is exceeded. The same applies to a greater extent to the polyols as the threshold dosage is significantly lower than for fructose. The results of animal experiments are thus in full accordance with the outcome of tests on human volunteers.

OSMOTIC DIARRHOEA

By means of the so-called saline laxative, a relatively uniform picture of acute diarrhoea can be obtained. As soon as the supply of this poorly absorbable substance is interrupted, the diarrhoea ceases.

Epsom salt (magnesium sulphate) takes effect primarily by way of its combination with water in the intestinal lumen, as a result of insufficient absorption. A slight absorption of magnesium sulphate does, of course, take place nevertheless. On the other hand, other salts regarded as easily absorbable can similarly bring on a characteristic osmotic diarrhoea, if they are administered in isolation and in large enough quantities. A typical example is sodium chloride (common salt), which, when administered in dosages of 20–40 g (0·3–0·6 mol) on an 'empty' stomach, also brings on an osmotic diarrhoea, even though both ions in themselves can easily be absorbed from the small intestine. Glauber's salt (sodium sulphate) must be assigned to an intermediate position. The cation, sodium, is quite easily absorbed, whereas the anion, sulphate, is not—or rather, only slightly—absorbed. As no free acid can originate within the intestinal lumen, the full osmotic effect of the sulphate (i.e. of cation and anion) is unchecked. The absorbable cation is 'bound' by the non-absorbable anion and is thus prevented from being absorbed, in the same way that water can be 'bound' osmotically by the non-absorbable, osmotically active substances in the intestine.

Results with Polyols

In evaluating the polyols with regard to their laxative effect, various phenomena may be observed. First, there is a close relationship with the osmotically active laxatives. Typical osmotic diarrhoea is brought on in a causal relationship by the combination of water in the intestinal region and without irritation of the mucous membrane. Similarly, the polyols are no longer sufficiently absorbed from the small intestine when administered in relatively high dosages, especially in an isolated form in drinks, and thus reach the region of the large intestine. Here their laxative effect can, in certain circumstances through the influence of bacteria, be significantly increased (in contrast with saline laxatives). The result of this bacterial influence is the metabolic formation from sorbitol and, admittedly to a lesser extent from xylitol, of decomposition products with a lower molecular weight and, therefore, with a greater osmotic effectiveness. As a rule, the products are organic acids (e.g. lactic acid, acetic acid, propionic acid, butyric acid), which are completely ionised by the pH of the large intestine as a result of their low pK value (pK around 3·5–5). This means that, in fact, the completely ionised salts of these organic acids are present. By the combination of an easily absorbable cation with an anion which can no longer be absorbed in the large intestine, the osmotic effect is once again increased. Calculations from a model based on the formation of lactic acid from sorbitol give the following picture: From 1 mol of sorbitol, 2 mol of lactic acid are formed, which are able to bind (to lactate) 2 mol of cations. The osmotic effect (and thus the laxative effect) of the sorbitol reaching the large intestine is thereby quadrupled! This applies similarly to any other substances which can be metabolised by the bacteria of the large intestine to form organic acids and which cannot be easily absorbed in the small intestine, provided they are administered in a sufficiently large quantity.

Typical examples of an extreme nature are osmotic diarrhoea in cases of poor digestion and poor absorption of disaccharides and monosaccharides. Indeed, in cases of inherited malabsorption of glucose/galactose, administration of these monosaccharides can bring on an extremely pronounced osmotic diarrhoea whereby the 'amplification effect' of bacterial influence in the large intestine is particularly significant (Meeuwisse and Melin, 1969). In the case of poor digestion of disaccharides (e.g. malabsorption of lactose or saccharose–isomaltose), the same effect is caused by the appropriate disaccharides in relatively small amounts (Simons, 1969). Besides absorbability, therefore, a precondition for the extent of the osmotic effect of organic substances is also the ease with which they can be metabolised into organic acids by the bacteria of the large intestine. In cases of lactose intolerance (lactose being the disaccharide of milk), $\frac{1}{4}$–$\frac{1}{2}$ litre of cow's milk (containing 7–14 g of lactose) can induce diarrhoea. As lactose intolerance occurs in large groups of the world's adult population on account of poverty (e.g. Thailand, Vietnam and in Hammitic tribes in Africa), milk is used in those regions as a laxative, whereas milk is seen as an exceptionally high-grade food by the lactose-tolerant peoples of Central Europe (Simons, 1969). Even in cases of lactose intolerance, the very pronounced effect of very small quantities of the disaccharide is explicable primarily on account of the bacterial 'amplification effect' The osmotic effect of the disaccharide is further limited by its high molecular weight. One osmole of lactose can yield 8 osmole of lactate!

In order to estimate correctly the theoretical preconditions for the possibility of osmotic diarrhoea being caused by certain substances, one must have regard not only to one's knowledge of the capacity of the small intestine for absorption (and digestion), but also to the possibility of bacterial decomposition in the large intestine. Lactose, sucrose and particularly glucose can be readily utilised in all cases by intestinal bacteria. In cases of inherited malabsorption or maldigestion of these sugars, the effects are correspondingly pronounced (Meeuwisse and Melin, 1969; Simons, 1969). In contrast, the capacity of bacteria to convert polyols is certainly lower. As far as we can see, sorbitol is better metabolised by intestinal microorganisms than xylitol. Additionally, xylitol has a lower molecular weight, which results in increased absorption.

As a result of low intestinal absorption in the case of sorbitol, exceeding the threshold value brings on an osmotic diarrhoea. This threshold value varies with each individual, but lies between a single dose of 8–10 g and a single dose of 15–30 g (Förster, 1972; Förster and Ziege, 1973). For xylitol, the threshold value at which diarrhoea is induced lies much higher, which may be due in addition to a lack of, or a reduction in, the bacterial decomposition. Most human subjects were able to tolerate without problems 30 g of xylitol in a single dose (Asano, Levitt and Goetz, 1972; Förster, 1972). It is important to mention again that the effect with preparations in liquid form (e.g. drinks) must be significantly stronger than when the substance is used in solids (e.g. in chocolate or in puddings). There are various reasons for this. Emptying of the stomach is slowed down by substantial meals and, therefore, smaller amounts of polyols reach the small intestine. The osmotically induced stimulation of peristalsis is reduced and passage through the small intestine is lengthened accordingly. For any given period of time, therefore, less polyols reach the small intestine, making possible a higher rate of absorption, and relatively fewer osmotically effective substances reach the large intestine.

Investigations using concentrated solutions of organic osmotic laxatives in water (e.g. as an experimental drink) are therefore not proof of the actual tolerance when the same substances are included in food, i.e. when they are administered in solids. They can only provide pointers to help in the choice of a suitable comparative substance. When using polyols as sweeteners, therefore, preparations in the form of drinks should be evaluated with much greater caution than when they are used in jams, for example, or even in sweets, as the possible and probable rate of intake is significantly thereby reduced.

EVALUATION OF EFFECTS UNDER THE CONDITIONS OF NUTRITION

For physiological requirements, sugar and sugar substitutes are used for sweetening, i.e. these substances are taken, as a rule, in combination with solids. The only exceptions are sweetened drinks; Coca-Cola and other commercial drinks contain up to 10–12% sucrose. In order to achieve a corresponding degree of sweetness and body, it would certainly be necessary to use more than 5% xylitol. If exclusive use were made of sorbitol, more than 15% would be necessary. It is quite possible and conceivable to consume 1 litre of liquid in a relatively short time (i.e. within 15–30 minutes). This means that, considering the lower concentration limit of 5%, it would be possible to consume 50 g of xylitol in drinks within a short time. Such a quantity would very probably lead to osmotic diarrhoea, judging by experiments with volunteer subjects. On the basis of the studies in Turku alone, even larger quantities of drinks containing xylitol were tolerated by people adapted to it. (Mäkinen and Scheinin, 1975; Scheinin, personal communication). However, it is not possible to generalise on the basis of these findings. For use in sweets, about 20–30 g of xylitol would be necessary to induce diarrhoea. As a normal sweet (or piece of chewing gum) contains about 1·5 g to 2·5 g of sugar, at least 5–10 sweets could be consumed before diarrhoea occurred. As far as the sweetening of foods goes, the conditions are more favourable still. In such cases the concentration of sugar or sugar substitute is relatively low, as more additional constituents are present. Overall, the emptying of the stomach is thereby retarded still further, and the dosage necessary to bring on diarrhoea is significantly higher as already explained.

These considerations of the basic physiological mechanism enable the experimental data to be explained. Complete absorption of polyols from the intestinal lumen, such as might be expected for, say, glucose or galactose, is rather improbable (Förster, 1962). The same applies in particular when polyols are used in greater concentrations to sweeten drinks. Complete absorption is more possible when smaller quantities of polyols are used to sweeten foods or when they are consumed in the form of sweets. Theoretical consideration and experimental data enable us to predict, in addition, that xylitol, because of its low molecular weight, will be absorbed much more readily than sorbitol.

It was possible to confirm these expectations in a feeding experiment carried out recently. Whereas an addition of 20 % xylitol to the food was partially tolerated by rats, distinct problems occurred with an addition of 10 % sorbitol. With an addition of 20 % sorbitol, many animals died (Förster, Heller and Fortmeyer, 1975; Heller, 1976). Dissection of the animals disclosed a swollen, gas-filled large intestine. This confirmed earlier results when using partly excessive dosages of xylitol (Kieckebusch *et al.*, 1961). In other feeding experiments, the good tolerance of xylitol was also established, as long as threshold dosages which could induce diarrhoea were not exceeded (Brin and Miller, 1974; Mosinger, 1971; Swarm and Banzinger, 1970).

In animal feeding experiments it was initially established that a sort of adaptation to xylitol occurred (Bässler, Stein and Belzer, 1966). The experimental animals were able to tolerate significantly greater amounts of xylitol when the dose rate was gradually increased, than when a large quantity of xylitol was administered from the outset. This 'adaptation' was thought to be associated with an enzyme induction (Kieckebusch *et al.*, 1961; Bässler *et al.*, 1962; Lang, 1971; Schmidt, Fingerhut and Lang, 1964). It was assumed that the enzyme in the liver which acts as a catalyst in the initial oxidation of xylitol (sorbitol dehydrogenase or xylitol dehydrogenase) is stimulated in its activity under the influence of constant administration of xylitol. A more rapid conversion of the xylitol absorbed from the intestine should thereby be achieved. By setting a higher level of concentration for xylitol its absorption from the intestinal lumen should be increased. This theory is not, however, necessarily in accordance with the true circumstances. Even with high oral dosages of polyols, their concentration in the peripheral blood hardly ever rose above 5 mg per 100 ml (Förster, unpublished).

It had already been indicated that when liquids are consumed, the polyol concentration in the intestinal lumen amounts to 2000–4000 mg per 100 ml. Even when relatively large amounts of polyols are administered with the feed, the concentration in the intestinal lumen may reach at least 100–400 mg per 100 ml. Concentrations of polyols in the portal vein, on the other hand, hardly ever exceed values of 20–30 mg per 100 ml when conditions are favourable (e.g. when the substance is administered directly into the intestine in high concentrations). Under conditions of oral administration these values will certainly be lower by a clear margin. For these reasons, therefore, no significant change in the level of concentration is to be achieved by a slightly accelerated hepatic uptake of xylitol.

INVESTIGATION OF THE TOLERANCE OF XYLITOL IN MAN

Experiments involving the oral administration of polyols were also carried out on man: those involving comparison with another sugar, e.g. glucose or fructose, are, of course, always the most significant. It had already been indicated that glucose has no laxative effect even in extraordinarily high dosages (Förster, Haslbeck and Mehnert, 1972). For fructose, the threshold dose lies around 70–100 g. Sorbitol at a single dose of only 15–30 g leads to diarrhoea in young subjects whereas approximately twice this quantity of xylitol (25–40 g) would be required to have a similar effect (Asano *et al.*, 1972; Förster, 1972; Förster and Ziege, 1973). In the early experiments performed by Mellinghoff (1961) xylitol was used as a substitute for sugar with diabetics. Using low dosages (10 g per day) there was hardly any symptoms of diarrhoea. Only when dosages were higher (3 × 20 g in tea), were there cases of diarrhoea among the diabetics. Similarly with metabolically healthy subjects, diarrhoea occurred after 2 × 25 g of xylitol in small quantities of liquid.

In the experiments performed by Dubach, Feiner and Forgó (1969), the effects of adaptation were investigated more closely in the human subjects. All subjects tolerated 40 g of xylitol distributed during the day. In this way three of the subjects were able to consume 200 g of xylitol or more per day. Some of the subjects suffered diarrhoea, eructation and flatulence. In our own experiments, 100 g of xylitol was tolerated without much difficulty by six volunteers over a period of 10 days (Förster, Boecker and Ziege, 1972). In the experiments performed by Mertz *et al.*, (1972) there were likewise no problems with an admittedly lower daily dosage of 50 g. It was possible to administer 30 g of xylitol per day over a period of 4 weeks to diabetic children (Förster, Boecker and Walther, 1977). Only one child withdrew prematurely from the sequence of experiments on account of diarrhoea. Xylitol was well tolerated by the other children as a substitute for sugar. In comparison, earlier experiments showed the tolerance of sorbitol by children to be not very good (Grybowski, 1966). In our own studies, the administration of 30 g of sorbitol to diabetic children had to be discontinued prematurely because of associated unfavourable reactions. In the studies in Turku, approximately 50 g of xylitol per day were consumed without ill effects over a number of years (Mäkinen and Scheinin, 1975). The maximum intake per day is reported to have been 200–400 g. There is, however, no clear-cut documentation for these figures, as they are based on observations and on retrospective calculations.

It has not been settled what proportion of orally administered xylitol is absorbed from the intestinal lumen in each particular set of circumstances—direct oral intake/consumption in combination with solids. It can, nevertheless, be regarded as established that even when administered in the most direct form in solution, a significant part of the xylitol must have been absorbed, as the associated metabolic effects can only be explained in this way, e.g. the increase in concentration of uric acid (Förster and Ziege, 1973), or the

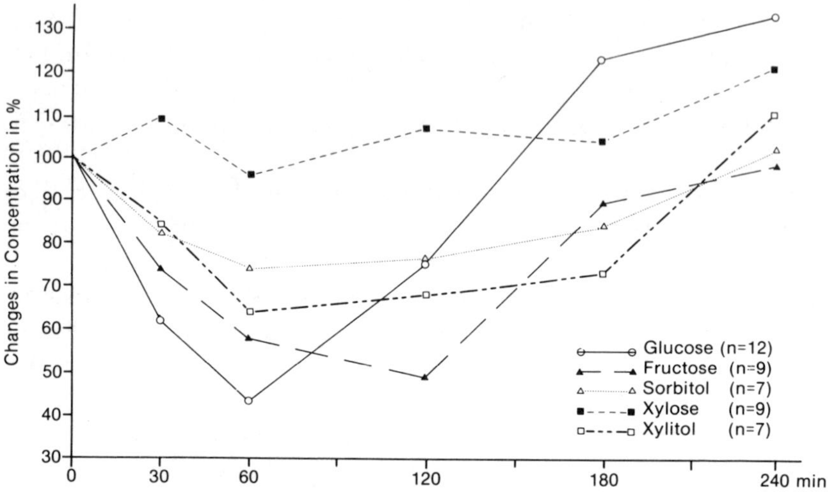

FIG. 10. Influence of oral supply of different carbohydrates on the concentration of free fatty acids in volunteer experimental subjects.

decrease in the concentration of fatty acid (Förster, 1972) (Fig. 10). This decrease in fatty acid concentration is a metabolic effect of the substances absorbed from the intestinal lumen. Xylose, for example, has no effect on fatty acid concentration although it is absorbed from the intestinal lumen.

CONCLUSIONS

After oral administration, xylitol is absorbed from the intestinal lumen by means of free diffusion. There is no indication of the presence of transport mechanisms of any sort. The motive force for the absorption of xylitol from the intestinal lumen is the level of concentration. When the capacity for absorption is exceeded as a result of excess dosage, it leads to osmotically induced diarrhoea. A healthy person can probably tolerate 10–30 g of xylitol in a single dose without ill effects and without diarrhoea. Likewise, in children, no negative effects occur with a single dose of 10 g of xylitol. Adult patients may be able to tolerate in all more than 200 g of xylitol per day after adaptation on graduated dosages. However, such cases are exceptional. As a rule, only 50–70 g of xylitol are tolerated by all adults without undue problems. The tolerance to xylitol is better when consumed together with other nourishment in the form of meals. Even when used in confectionery (sweets, chocolate, chewing gum) the danger of diarrhoea is not very great, because in general only small quantities are consumed. On the other hand, when used as a sweetener in lemonade or other soft drinks, there is the possibility that xylitol will give rise to diarrhoea, as in this case the intake will be relatively high in a short time. These considerations do not apply to the same extent for the use of xylitol in tea or coffee, as the amount consumed will be measured directly and can be limited (e.g. to 10 g for each single dose). It should be mentioned that when dietetic use is made of relatively large quantities of xylitol or sorbitol, the use of laxatives should either be discontinued completely, or the need for laxatives should be reconsidered.

REFERENCES

Asano, T., Levitt, M. & Goetz, F. C. (1972) *Diabetes*, **21,** 350.
Bässler, K. H., Prellwitz, W., Unbehaun, V. & Lang, K. (1962) *Klin. Wschr.*, **40,** 791.
Bässler, K. H., Stein, G. & Belzer, W. (1966) *Biochem. Z.*, **346,** 171.
Bogner, P. H. (1961) *Proc. Soc. Exp. Biol. Med.*, **107,** 263.
Bogner, P. H. & Haines, I. A. (1961) *Proc. Soc. Exp. Biol. Med.*, **107,** 265.
Brin, M. & Miller, O. N. (1974) The safety of oral xylitol. In: *Sugars in Nutrition*, eds. H. L. Sipple & K. W. McNutt, p. 591. New York: Academic Press.
Cori, C. F. (1925) *J. Biol. Chem.*, **66,** 691.
Csaky, T. Z. (ed.) (1975) *Intestinal Absorption and Malabsorption*. New York: Raven Press.
Dubach, K. C., Feiner, E. & Forgó, I. (1969) *Schweiz. med. Wschr.*, **99,** 190.
Förster, H. (1962) *Inaugural Dissertation*, München.
Förster, H. (1972) *Med. Ernähr.*, **13,** 7.
Förster, H. & Menzel, B. (1972) *Z. Ernährungswiss*, **11,** 24.
Förster, H. & Ziege, M. (1973) *Z. Ernährungswiss*, **12,** 211.
Förster, H., Boecker, S. & Walther, A. (1977) *Fortschr. Med.*, **95,** 99.
Förster, H., Boecker, S. & Ziege, M. (1972) *Med. Ernähr.*, **13,** 193.

Förster, H., Haslbeck, M. & Mehnert, H. (1972) *Diabetes*, **21,** 1102.
Förster, H., Heller, G. & Fortmeyer, M. P. (1975) *Infusionstherapie*, **2,** 327.
Förster, H., Meyer, E. & Ziege, M. (1972) *Rev. Europ. Clin. Biol.*, **10,** 958.
Grybowski, D. (1966) *New Engl. J. Med.*, **275,** 718.
Guy, M. I. & Deren, J. J. (1971) *Amer. J. Physiol.*, **221,** 1051.
Haslbeck, M. (1976) *Habilitationsschrift*, München.
Heller, G. (1976) *Inaugural Dissertation*, Frankfurt.
Holdsworth, C. D. & Dawson, A. M. (1964) *Clin. Sci.*, **27,** 371.
Kieckebusch, W., Gziem, W. & Lang, K. (1961) *Klin. Wschr.*, **39,** 447.
Lang, K. (1971) *Med. Wschr.*, **49,** 233.
Liebau, K. (1964) *Inaugural Dissertation*, München.
Mäkinen, K. K. & Scheinin, A. (1975) Turku sugar studies. VI. The administration of the trial and the control of the dietary regimen. *Acta Odont. Scand.*, **33,** Suppl. 70, 105.
Meeuwisse, G. & Melin, K. (1969) *Acta Paediat. Scand. Suppl.*, 188.
Mehnert, H. & Förster, H. (1968) *Diabetologia*, **4,** 26.
Mehnert, H., Stuhlfauth, K., Mehnert, B., Lausch, R. & Seitz, W. (1959) *Klin. Wschr.*, **37,** 1138.
Mellinghoff, C. H. (1961) *Klin. Wschr.*, **39,** 447.
Mertz, D. P., Kaiser, V., Klöpfer-Zaar, M. & Beisbarth, H. (1972) *Klin. Wschr.*, **50,** 1107.
Mosinger, M. (1971) Cited by Brin, M. & Miller, O. N. (1974).
Packer, L. (1974) *Biomembranes*. New York: Academic Press.
Schmidt, B., Fingerhut, M. & Lang, K. (1964) *Klin. Wschr.*, **42,** 1073.
Simons, F. J. (1969) *Amer. J. Digest. Dis.*, 14.
Swarm, R. L. & Banzinger, R. (1970) Cited by Brin, M. & Miller, O. N. (1974).
Verzar, F. (1935) *Biochem. Z.*, **276,** 17.
Verzar, F. & Laszt, L. (1935) *Biochem. Z.*, **276,** 17, 28.
Wilson, T. H. (1962) *Intestinal Absorption*, Philadelphia: Saunders.
Wright, E. M. (1974) Passive permeability of the small intestine. In: *Intestinal Absorption*, ed. D. H. Smyth, p. 159. London: Plenum Press.

CHAPTER 6

Biochemical Findings in Exocrine Secretions in Relation to Peroral Administration of Xylitol

K. K. MÄKINEN
(*Institute of Dentistry, University of Turku, Turku, Finland*)

INTRODUCTION

The chemical composition of the diet affects the biochemical properties of the exocrine secretions involved in digestion. This is natural and expected. Consequently, quite considerable variations in the chemical quality of saliva, pancreatic juice and other secretions can be found between various subjects and geographical areas without such variations being considered abnormal in any way. So, for example, the concentration of the salivary α-amylase depends on the relative proportion of hexose-based carbohydrates in the diet. Specific dietary sugars seem to control certain enzyme activity in the human small intestine as well. The ingestion of sucrose instead of starch by rats alters the levels of various enzymes in the liver but glucose and fructose do not have a similar effect (Basu, Dickerson and Parke, 1975). Recent findings made in this laboratory suggest that dietary carbohydrates may also influence the function of exocrine glands which are not involved in digestion. Virtually all available information about the effects of dietary sugars on exocrine gland function has been obtained with glucose, fructose, common disaccharides, starch and related carbohydrates, while the study of the effects of sugar alcohols on exocrine glands, apart from the pancreas, has only recently been started. The knowledge about these effects is thus scant, but certain indicative results have already been obtained.

XYLITOL-ASSOCIATED CHANGES IN SALIVARY LACTOPEROXIDASE LEVELS

According to a broad definition, lactoperoxidase (donor:H_2O_2 oxidoreductase, EC 1.11.1.7) catalyses the following general reaction:

$$\text{Donor} + H_2O_2 = \text{oxidised donor} + 2H_2O$$

Probably the first observation of the nexus between the consumption of xylitol and increased salivary lactoperoxidase activity was made in connection with the Turku sugar studies (Mäkinen, Tenovuo and Scheinin, 1975). In this study, pooled whole saliva samples were analysed for lactoperoxidase at the 3·5, 7·5, 13, 20 and 24 month stages of the 2-year feeding trial. At every analysis point the saliva pools of the xylitol group displayed the highest specific peroxidase activity (Fig. 1).

Whole saliva also contains pseudoperoxidases and peroxidases of bacterial and leucocyte origin. The separation of these enzymes was, however, easy to perform using chromatographic methods such as carboxymethylcellulose chromatography (Fig. 2). The true salivary lactoperoxidase is bound to the cellulose and desorbed at low concentrations of NaCl. Figure 2A shows the enzyme activity expressed in seconds (the time required to reach a predetermined extinction in the reaction mixture involved), but the expression of the results as specific activities did not alter the results, as shown in Fig. 2B (Mäkinen, Tenovuo and Scheinin, 1976). It was subsequently suggested that the xylitol-induced elevation of the salivary lactoperoxidase activity and the cariostatic properties of xylitol are partly inter-related phenomena. The above human study needed verification particularly as it had not been possible to take into account certain important variables, such as the flow rate of saliva. Furthermore, the presence of bacteria in the samples, until they were spun down, was also a matter to be considered. The verification was

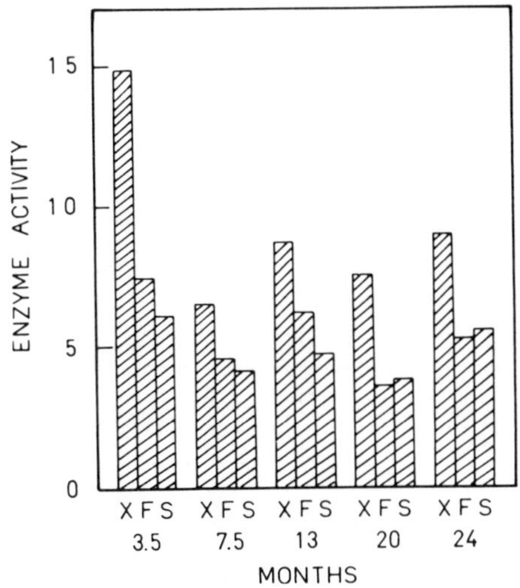

FIG. 1. Specific lactoperoxidase activity of the supernatant fluid (23 500 × *g*) of human whole saliva. Saliva samples were collected at the times indicated from subjects on a strict diet with regard to the sweeteners used (xylitol, X; fructose, F; sucrose, S). Carboxymethylcellulose chromatography showed that the bulk of the peroxidase activity in the xylitol and fructose groups was caused by the salivary lactoperoxidase, whereas in the sucrose group the contribution of microbial and other peroxidases was considerable. (Mäkinen, Tenovuo and Scheinin, 1975.)

carried out on monkeys (*Macaca mulatta*) which were fed a xylitol-containing diet for three days. Samples of parotid and submandibular saliva were collected on the second and third day and they were analysed for lactoperoxidase activity. Ingestion of xylitol was clearly associated with a significant increase of the activity of the salivary lactoperoxidase, no matter in which way the results were expressed (U/ml, U/mg or U/min). The increase could be observed in both parotid and submandibular saliva. The concentration of protein was also increased, as shown with two separate methods, the Lowry colorimetric method (Mäkinen *et al.*, 1977A) and by determining the absorbance of diluted saliva at 215 nm (Bird *et al.*, 1977). The results from both the human and the monkey studies thus indicated that at the moment when the collection of saliva was started, the acinar units of the salivary glands were filled with fluid containing proteins and enzymes, the concentrations of which

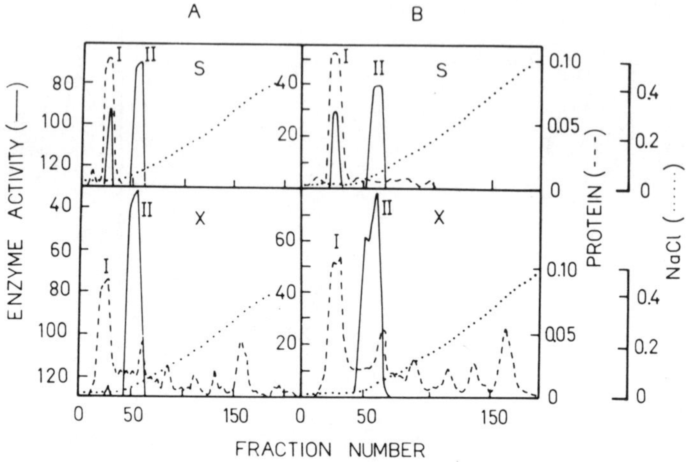

FIG. 2. Separation of salivary and other peroxidases from human whole saliva on carboxymethylcellulose columns. The left chromatograms (A) give the enzyme activity in seconds (the higher the activity, the more rapid the reaction) and the right ones (B) in specific activity (units/mg protein). I, unabsorbed peroxidases (bacterial peroxidases, myeloperoxidase, pseudoperoxidases and possibly aggregated lactoperoxidase); II, true salivary lactoperoxidase. S, sucrose; X, xylitol. (Mäkinen, Tenovuo and Scheinin, 1975 and 1976.)

depended on whether monkeys were fed xylitol or sucrose. One should not regard this type of phenomenon as harmful to the salivary glands or oral tissues.

Although this type of response of the salivary glands to dietary carbohydrates may have biological significance in general, the lactoperoxidase findings should especially be considered in the light of the proposed antibacterial rôle of this enzyme (Morrison and Steele, 1968). Lactoperoxidase is regarded as an enzyme which is partly responsible for the elimination of pathogenic bacteria from the lachrymal fluid, milk, saliva, oral and other mucous membranes. On the other hand, the Turku sugar studies showed that the consumption of a xylitol diet is associated with a strong reduction in the incidence of dental caries. This causal connection has led several scientists to believe that the xylitol-induced increase in the activity of lactoperoxidase would alone explain the advantageous dental effect of xylitol. The lactoperoxidase certainly plays a rôle in this context, but other factors should also be considered. Such factors include, for example, the non-fermentability of xylitol in human dental plaque and the effect of xylitol on the pH values of oral fluid.

The above two studies have led to numerous other experiments which have attempted to elucidate the nature of the lactoperoxidase effect. It was important that the lactoperoxidase effect was also demonstrated by an independent group (Harper, Poole and Wolf, 1977). In this case, the stimulation of parotid saliva with xylitol-sweetened fruit pastilles (containing gum arabic and sorbitol) increased the activity of the salivary lactoperoxidase in human subjects. Since the publication of their study, the authors have completed the experiments by increasing the number of subjects. Due to the high standard deviations involved in the determination of lactoperoxidase in human saliva, the original significant difference between xylitol and sucrose pastilles was changed to comprise a trend in favour of the xylitol product. The means differed, however, to such an extent that one is entitled to speak about three similar results of lactoperoxidase studies obtained so far.

As a direct continuation of the experiment of Harper *et al.* (1977), we carried out in this laboratory a similar series of tests on human parotid saliva stimulated with the above fruit pastilles. The experiments were, however, widened to comprise a whole 3-h period after the stimulation (Mäkinen *et al.*, 1977E). The subjects chewed 16 fruit pastilles (corresponding to 5 g xylitol) during a period of 15 min. Parotid saliva samples (3 × 5 min) were collected at 0, 1, 2 and 3 h directly from the opening of the duct allowing the excretion to run in a spoon. The stimulation of saliva for each 15-min collection period was performed by spraying every 5 min 1 % citric acid solution on the tongue. Consequently, in this study no cannulation or suction cups were necessary. The parotid saliva samples obtained in this way were considered quite suitable for all biochemical studies. The flow was brought about in a rather physiological manner. The samples were not allowed to come into contact with oral tissues other than the opening of the Stensen's duct. Altogether 312 parotid saliva samples were collected, one-half after xylitol stimulation and the other half after sucrose stimulation. The results showed that great variations occur between individuals in the

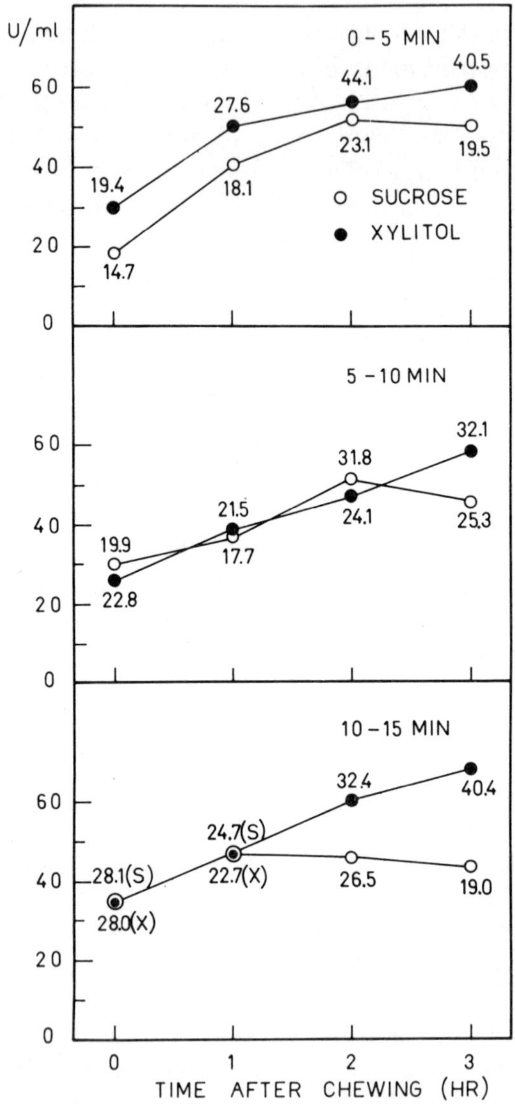

FIG. 3. Effect of chewing of sucrose- and xylitol-containing fruit pastilles on the lactoperoxidase activity of human parotid saliva ($n = 13$). Parotid saliva samples were collected as three separate aliquots during a period of 15 min, taking 5 min to collect each aliquot. The collections were carried out immediately after the 15-min chewing period (0) and 1, 2 and 3 h later. The dose was 5 g xylitol or sucrose. The standard deviations are indicated. (Mäkinen *et al.*, 1977E.)

activity of parotid saliva lactoperoxidase. A few samples did not show any detectable reaction at all, particularly those taken during the first 5-min period after sugar stimulation. High standard deviations are characteristic of many biochemical saliva parameters. Statistical analysis of the data indicated, however, that the chewing of xylitol fruit pastilles caused a higher lactoperoxidase activity than the corresponding sucrose pastilles (Fig. 3). The results were again expressed in U/ml, U/mg and U/min, and the effect of xylitol was more clearly seen in the U/ml and U/min versus time graphs. The protein concentrations are shown in Fig. 4. The differences in favour of xylitol were not significant in any of the cases shown in Fig. 3, but there was a trend of xylitol pastilles producing a higher lactoperoxidase activity 2–3 h after stimulation. The specific activity was not changed and the results can be explained as an increase of flow rate and protein concentration.

The data presented in Figs. 3 and 4 were obtained with 13 subjects. The study was completed by analysing four more subjects. The nature of the results did not change to any significant extent. Details of the investigation have been published separately (Mäkinen *et al.*, 1977E). In addition to the four lactoperoxidase studies mentioned above, which all yielded essentially similar results, there have also been other experiments in which the connection between xylitol and lactoperoxidase activity has been investigated. Table 1 presents a summary of all experiments known to the author at the present time. Examination of the experimental conditions and the data obtained reveal, among other things, the following:

1. A positive lactoperoxidase effect (i.e. an increase) seems to be associated with prolonged use (at least 2 or 3 days) or even with a single dose, provided that in the latter case a sufficiently long mastication or sucking period (15 min) has preceded the sample collection which should be carried out 2–3 h following stimulation.
2. Mere use of one piece of xylitol-containing chewing gum does not affect the activity of lactoperoxidase to any noticeable extent. Consequently, the effect also seems to be concentration dependent.

TABLE 1
List of Studies on the Connection Between Xylitol and Salivary Lactoperoxidase
The list includes experiments in which xylitol was administered enterally

Subject	*Type of saliva studied*	*Length of administration*	*Experimental details*	*Effect on lactoperoxidase*	*Authors*
Man (Turku sugar study)	Mixed (oral fluid)	2 years	Paraffin stimulation. Total substitution of sucrose with xylitol	Significant increase as compared to consumption of sucrose or fructose	Mäkinen, Tenovuo & Scheinin (1975, 1976)
Monkey (*Macaca mulatta*)	Submandibular and parotid	2–3 days	Pilocarpine stimulation. Total substitution. Collection by cannulation	Significant increase as compared to sucrose consumption	Mäkinen *et al.* (1977A)
Man	Parotid	Single dosage	7–16 fruit pastilles sweetened with a mixture of xylitol and sorbitol, or sucrose	Indicative increase in the xylitol group	Harper, Poole & Wolf (1977)
Man	Parotid	Single dosage	16 fruit pastilles (no sorbitol); samples collected immediately after pastilles and 1, 2 and 3 h later	Indicative increase in activity after xylitol in 2 and 3 h samples	Mäkinen *et al.* (1977E)

Man	Mixed (oral fluid	One year	Paraffin stimulation. Partial substitution using chewing gum only (6·7 g/day) not immediately before sample collection	No change in enzyme level	Mäkinen & Scheinin (1975C)
Man	Mixed (oral fluid)	Single dosage	Paraffin stimulation; 0·6 g/kg xylitol or glucose. 200 ml solution in 1 min	No change as compared to glucose over a period of 6 h after sugars	Mäkinen *et al.* (1977C)
Man	Mixed (oral fluid)	Single dosage	Rinse with 10 % solution (15 ml) or chewing one stick of chewing gum (1·5 g sugar per stick). Collection by paraffin stimulation	No significant change in enzyme levels over a period of 30 min after sugars	The present article
Man	Mixed (oral fluid)	One month	Chewing of chewing gum sweetened with various mixtures of sorbitol and xylitol (8 sticks per day). The stimulation at the start was performed with paraffin and one month later with the polyol gums	No significant change between various mixtures of sugar alcohols, but the levels of lactoperoxidase were lower after 1 month intensive chewing than before the experimental period	Mäkinen *et al.* (1977B)

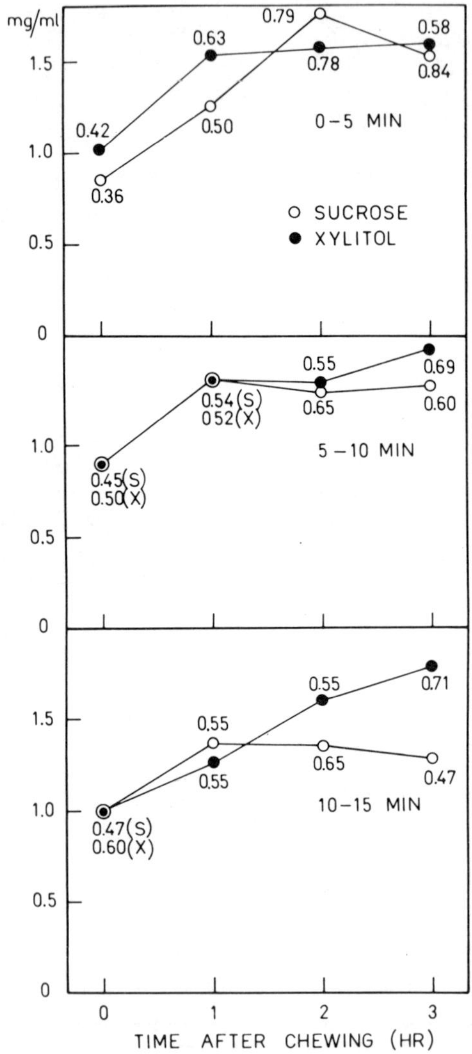

FIG. 4. Effect of chewing of xylitol- or sucrose-containing fruit pastilles on the concentration of proteins in human parotid saliva. All details are as for Fig. 3.

3. A single mouth rinse with 10% xylitol solution may not produce a positive effect.

One should also consider the possibility that the activity of lactoperoxidase *per se* in whole saliva may not be a decisive factor; it is probable that the concentration of the final inhibitory compound, the hypothiocyanate ion, $OSCN^-$, formed by hydrogen peroxide and SCN^- ions, may be critical (Hoogendoorn *et al.*, 1977).

It is possible that under certain special dietary circumstances the levels of lactoperoxidase in whole saliva may significantly decrease. For example, when human subjects used approximately eight sticks of chewing gum per day over a period of four weeks, there was a significant reduction in the activity of lactoperoxidase both in stimulated and resting saliva (Fig. 5). The chewing gums were sweetened with various mixtures of xylitol and sorbitol. The exact reason for the decrease of the lactoperoxidase activity is not known, but it is conceivable that during intensive mastication the synthetic

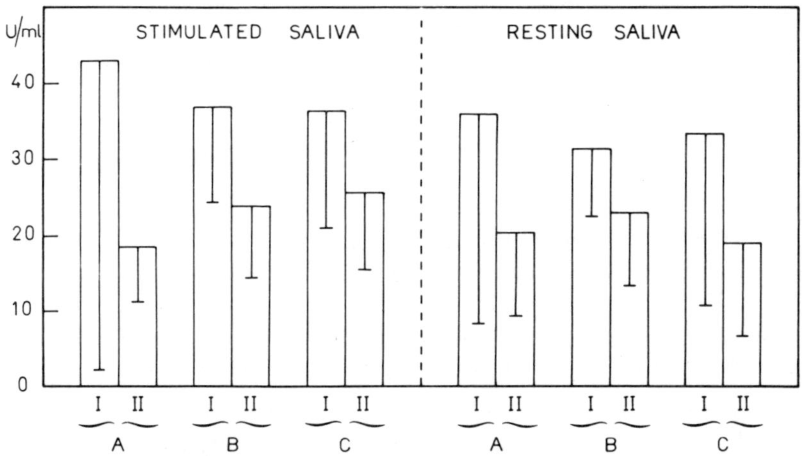

FIG. 5. Effect of chewing of polyol-containing chewing gums on the activity of lactoperoxidase of human whole saliva (supernatant fluid, 23 500 × *g*). The subjects in three groups (A, B, C) chewed approximately eight sticks of chewing gum per day. The chewing gums differed with regard to their xylitol:sorbitol ratio (w/w) as follows: A, no xylitol, 56% sorbitol; B, 7% xylitol and 49% sorbitol; C, 68% xylitol and 8% sorbitol. I, start of the study; II, after a 4-week chewing period. (Mäkinen *et al.*, 1977B.)

machinery responsible for the formation of the lactoperoxidase was not able to maintain a maximum enzyme level in saliva, although the protein synthesis itself most likely took place at a high rate. These effects seem to be selective as the secretion of sialic acids (or sialoproteins) was increased by intensive chewing of chewing gum, as shown in Fig. 6 (Mäkinen *et al.*, 1977B). The increase of the

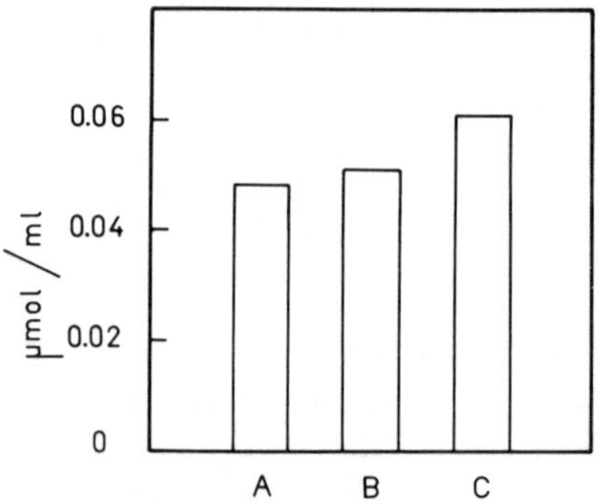

FIG. 6. Effect of various stimuli on the concentration of total sialic acids in human parotid saliva. Parotid saliva was stimulated either by paraffin (A), sucrose chewing gum (B) or xylitol chewing gum (C) for 2 min and parotid saliva was collected during a period of 15 min. (Unpublished results from this laboratory.)

concentration of sialic acid was accompanied by a similar increase of reducing sugars and viscosity of whole saliva.

The effect of mouth rinses with 10 % solutions of xylitol or sucrose was studied in this laboratory. The arrangements of the study are shown in Fig. 7. A 3·0 ml whole saliva sample was first collected at 8 a.m. using paraffin stimulation. The mouth was rinsed with 15 ml of a 10 % solution of either xylitol or sucrose (water with no sugar was used as a control) simultaneously chewing a piece of paraffin. No sample was collected during this 2 min rinsing/chewing period. Whole saliva samples were then collected during 2-min periods 2, 5, 10, 20 and 30 min after the sugar rinses. Consequently, the first 2-min sample was collected immediately after rinsing/chewing. The experiments were repeatedly carried out on five human subjects (two to five times). The sugar solutions were prepared on a percentage, not on a molar basis, as xylitol and sucrose, due to their equal

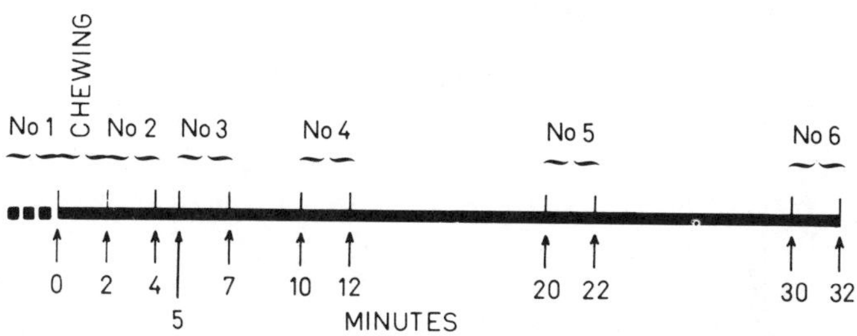

FIG. 7. The time schedule of a mouth-rinse chewing experiment on human subjects. A 3·0-ml whole saliva sample (No. 1) was collected immediately before a 2-min chewing/rinsing period. The rest of the samples (Nos. 2–6) were subsequently collected. The chewing/rinsing comprised chewing of a 1-cm^3 piece of paraffin with simultaneous rinsing with 15 ml of a 10 % (w/v) sugar solution (water without sugars as the control). All samples were collected by paraffin stimulation. The flow rates were recorded and the samples were immediately analysed for pH, lactoperoxidase and other parameters.

sweetness, can generally be used in equal amounts in commercial products. All saliva samples were collected by paraffin chewing (a new piece for each collection) and the flow rate was recorded. The activities of peroxidase in the samples expressed either as U/ml or U/mg protein are shown in Figs. 8 and 9. No significant differences between the use of water, xylitol or sucrose solution were found. As usual, the standard deviations were considerable. Figure 10 shows the protein concentration of the whole saliva samples studied. No significant differences between the three rinses were found. Although no significant differences in the lactoperoxidase activity were observed, such changes were, however, shown with regard to the pH values measured, as described later.

When the lactoperoxidase activity of human whole saliva was studied in pilot scale in samples obtained in a study similar to the above but using chewing gum instead of rinses, the results shown in Fig. 11 were obtained. In this case the graphs show the differences in the activity (in U/ml) calculated from the starting value determined on samples which were obtained immediately before chewing. All four types of chewing gum reduced the activity of peroxidase. Due to the preliminary nature of these results, the differences between sugars should not be considered very significant. The pH effects may

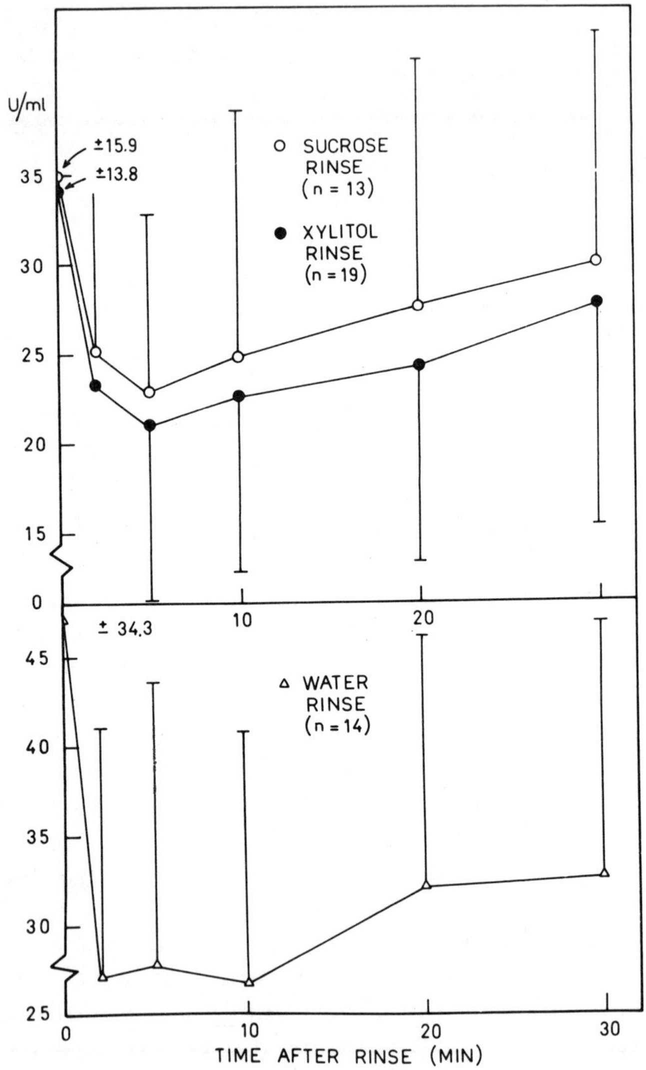

FIG. 8. Effect of xylitol and sucrose rinses on the activity (in U/ml) of lactoperoxidase of centrifuged (23 500 × *g*) human whole saliva Details are given in text and in Fig. 7.

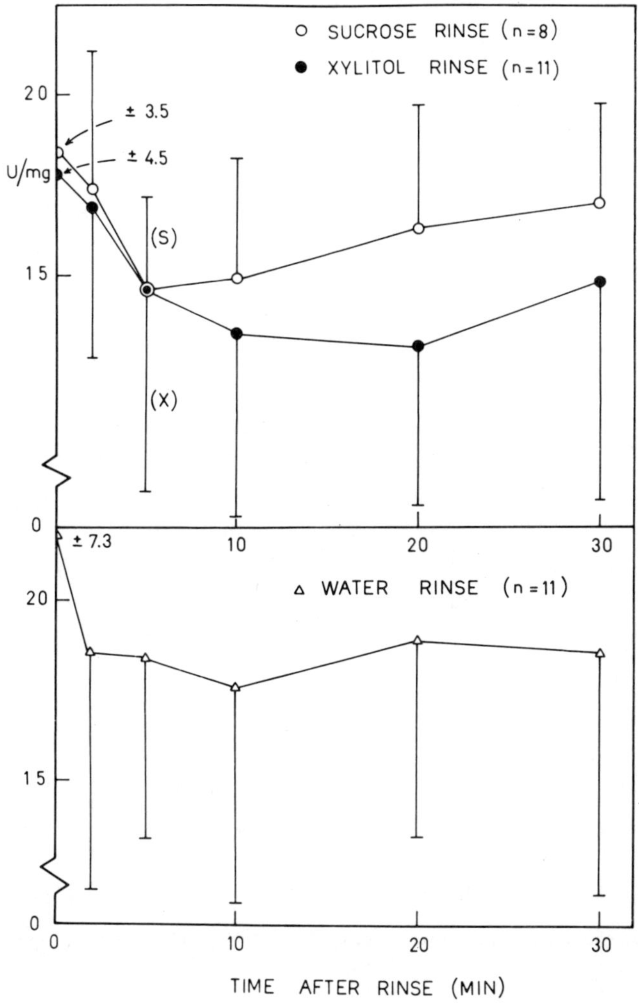

FIG. 9. Effect of xylitol and sucrose rinses on the activity (in U/mg) of lactoperoxidase of centrifuged human whole saliva Details are given in text and in Fig. 7.

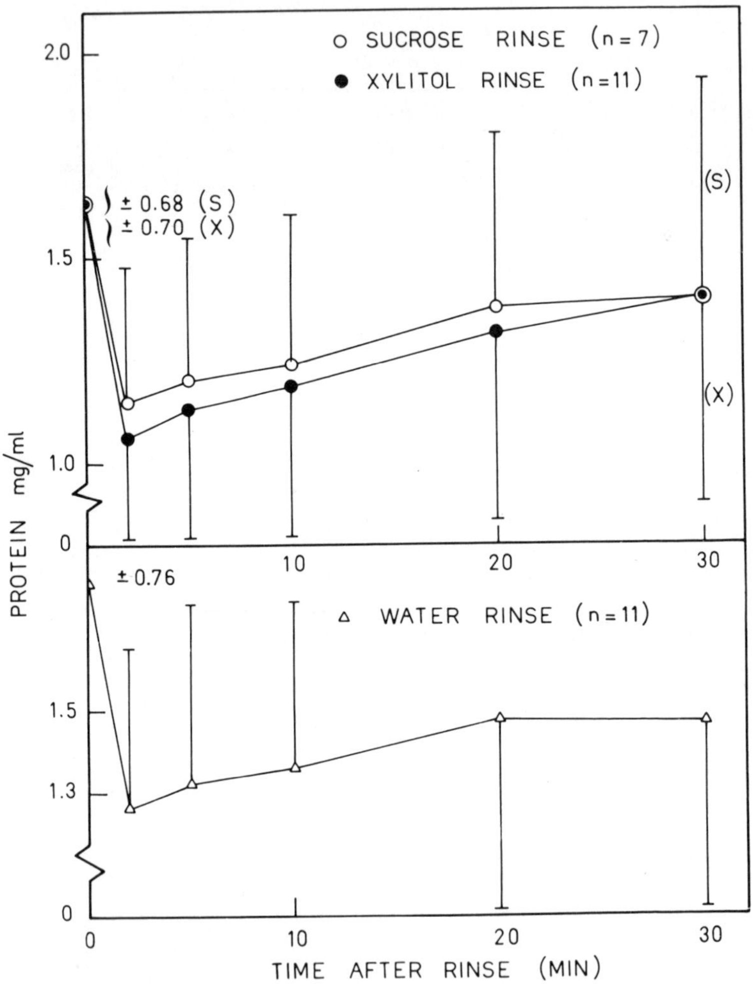

FIG. 10. The concentration of proteins in whole saliva of the experiment described in Fig. 7.

be compared as well (Fig.12). The results are shown as ΔpH, calculated from the starting values determined on samples collected before using chewing gum. There was a clear difference between the sugar alcohols sorbitol and xylitol on one hand, and the other two sugars (sucrose and fructose) on the other.

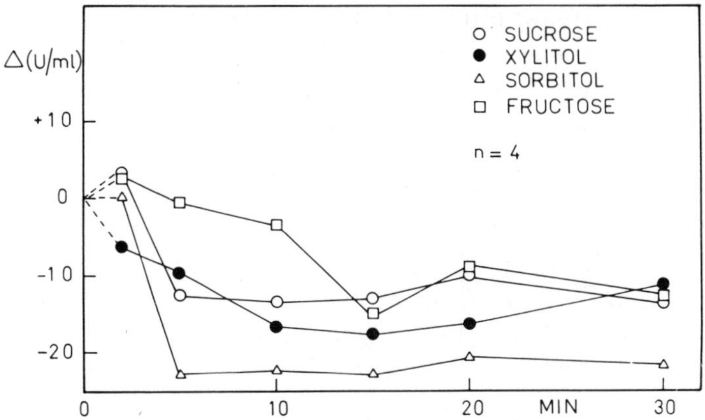

FIG. 11. Effect of chewing of chewing gums on the activity of lactoperoxidase of human whole saliva. Result from a pilot study ($n = 4$) similar to that described in Fig. 7, but replacing the sugar rinse with one piece of chewing gum sweetened with either sucrose, fructose, sorbitol or xylitol. No saliva samples were collected during the 2-min chewing period. All samples were collected by paraffin stimulation. The results are expressed as activity differences compared to the activity determined for the sample collected before chewing. The gums contained 50% (w/w) of the sweeteners indicated.

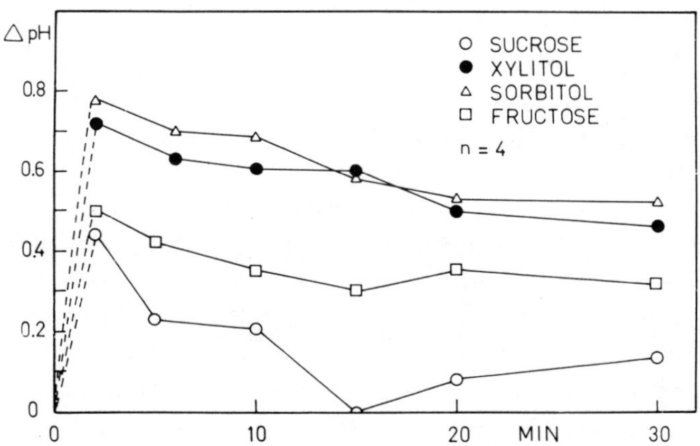

FIG. 12. Results of pH determinations of the experiment described in Fig. 11. The results are expressed as differences of pH compared to the initial pH values determined immediately before chewing of chewing gum.

There is very scant information about the effects of other sugars on the lactoperoxidase activity of human saliva. Preliminary experiments carried out in this laboratory suggest, however, that none of the sugars mentioned in Fig. 12 has any detectable effect on the activity of lactoperoxidase when used in 10 % rinse solutions in the way described above.

Greatly varying results have thus been obtained when testing the effects of various ways of administering xylitol on the salivary lactoperoxidase activity. The initial decrease of the activity after a single piece of chewing gum or a single mouth rinse should not be regarded as a harmful consequence. The changes in enzyme activity were similar no matter which sugar was used. The results were also similar to those obtained by rinsing with water alone. The salivary glands simply empty a part of their contents following a stimulus. This leads to a transitory decrease of certain salivary ingredients. The stimuli set protein synthesis going to cope with the increased need of enzymes and other molecules in saliva. The situation seems to be different when prolonged use of xylitol is involved, or when the dose, although single, is high enough or administered in a proper manner. A concomitant increase in the levels of lactoperoxidase in saliva has then been observed both in humans and monkeys.

The exact physiological and biochemical mechanism involved in the xylitol-induced increase of the lactoperoxidase activity is not known. A recent review article listed a number of possible reasons (Mäkinen, 1978). Against this background it is interesting to view the suggestion that thyroxine may act as a regulator of the iodinase and peroxidase enzyme(s) of the submaxillary gland (Chandra, Das and Datta, 1977). It was also suggested (Mäkinen, 1978) that the addition of an exogenous intermediate of the glucuronate cycle would result in a stimulation of the efficacy of the pathway, thereby possibly enhancing the synthesis of glycoproteins and/or mucopolysaccharides. As lactoperoxidase contains a carbohydrate moiety, such a possibility is attractive, but clearly confirming experiments should be carried out. Furthermore, it has not been possible to show with certainty, that enteral xylitol stimulates the glucuronate cycle. However, monkeys on a xylitol diet produce very viscous saliva (Mäkinen *et al.*, 1977A).

It has not been possible to explain the results only in terms of dehydration of the acinar units on xylitol diet. The clinical significance of the increase of lactoperoxidase activity would not, however, be exaggerated.

α-AMYLASE AND CONSUMPTION OF XYLITOL

The rôle of the pancreatic amylase (α-1,4-glucan-4-glucanohydrolase, EC 3.2.1.1) is to hydrolyse starch, amylodextrins and glycogen to maltose and some glucose. In general, the enzyme hydrolyses α-1,4-glucan links in polysaccharides containing three or more α-1,4-linked D-glucose units. The enzyme does not thus affect the points of branching in which the linkages are 1,6. Pancreatic amylase is thus an endoamylase which randomly hydrolyses the inner α-1,4-glycosidic linkages. The enzyme is secreted in the active form, but requires chloride ions for optimum activity. Pancreatic amylase appears to be identical with salivary amylase (Meyer *et al.*, 1948), although both are actually mixtures of amylases, each acting in its own particular manner. There are species differences in the amylase content of salivary glands. For diagnostic purposes it must be remembered that elevations in the blood level are transitory. The level usually remains high in acute conditions for 2–3 days and then returns to normal (in urine the level remains high for another 7–10 days). The properties of amylases have been briefly reviewed, for example, by Herman (1974).

It is conceivable that the amounts of amylase secreted from the pancreas depend on the concentration of substrate carbohydrates consumed. This is the case at least with certain salivary enzymes. On a predominantly carbohydrate diet, the concentration of salivary amylase is higher than on a low carbohydrate diet, as shown by Walker more than fifty years ago (Afonsky, 1961). Even rather heavy loading of human subjects with xylitol (approximately 70 g/day over a period of 2 years) did not affect the serum concentration of α-amylase to any noticeable extent (Fig. 13A). The difference between the sucrose and xylitol groups was almost significant but all values were within the physiological range. The data were obtained from one phase of the 2-year feeding study only, so that additional information on this matter should be obtained.

The determination of α-amylase activity in human whole saliva according to the standard serum assay is not fully reliable, as whole saliva contains some bacterial enzymes capable of hydrolysing

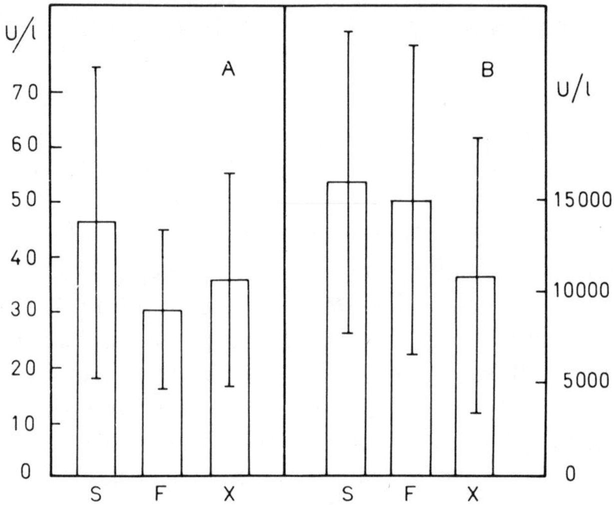

FIG. 13. Effect of chronic consumption of xylitol, fructose or sucrose on serum (A) and whole saliva (B) α-amylase activity. The data are from the 2-year feeding study, A from the 13·5 month and B from the 24 month stage. (Mäkinen and Scheinin, 1975A and 1975B.)

starch (Mäkinen and Scheinin, 1975A). The α-amylase activity of the parotid secretion is, however, so high that the possible microbial enzymes are generally masked. Consequently, although some microbial enzymes may be involved, the result shown in Fig. 13B for whole saliva starch-splitting enzyme activity also gives an idea of the levels of the true α-amylase in whole saliva of subjects consuming xylitol, fructose or sucrose. The response of the human salivary α-amylase to sugars needs, however, experimental confirmation. This is necessary because the consumption of a xylitol diet significantly increased the α-amylase activity of monkey parotid saliva, expressed in Somogyi units (Bird *et al.*, 1977). The content of total protein was simultaneously increased as previously mentioned, but the specific

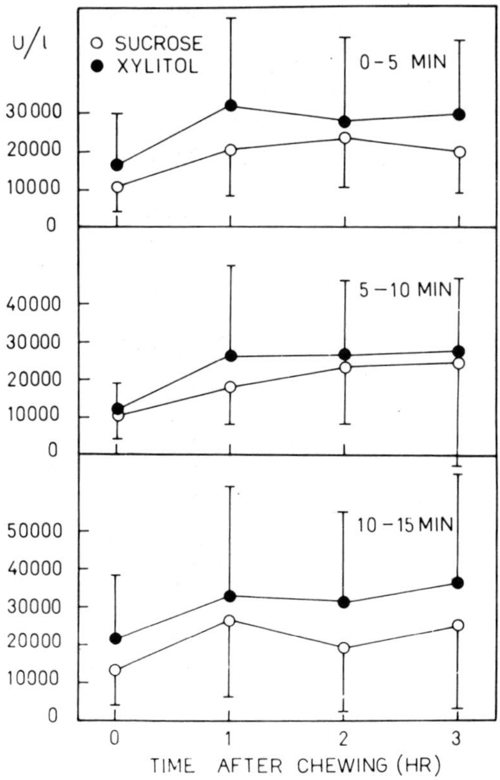

FIG. 14. Effect of chewing of xylitol- or sucrose-containing fruit pastilles on the activity of α-amylase of human parotid saliva. Details as for Fig. 3. (Mäkinen *et al.*, 1977E.)

activity of amylase expressed in units/mg did not differ between sucrose and xylitol diets. The specific α-amylase activity thus behaved in a fully opposite manner when compared with the specific lactoperoxidase activity. α-Amylase forms, however, a considerable portion of the proteins in saliva, whereas the molar concentration of lactoperoxidase is very small. Hence changes in the specific lactoperoxidase activity by sugar diets may be more readily observable than in the case of α-amylase.

No pathological changes in the pancreas have been detected in connection with oral administration of xylitol in man. Prolonged consumption with high doses may to a certain extent be reflected in lowered levels of the enzyme in the serum (Mäkinen & Scheinin, 1975B), but this type of change is understandable, and expected within the physiological range. A predominantly monosaccharide diet, particularly if it is not based on glucose units, may result in this sort of pancreatic response. Single doses of 0·6 g/kg were not able to evoke any differences between glucose and xylitol, however (Mäkinen *et al.*, 1977C).

Preliminary experiments on human parotid saliva showed that chewing of 16 fruit pastilles during a period of 15 min caused a small increase in the salivary α-amylase activity when compared with the effect caused by sucrose pastilles (Mäkinen *et al.*, 1977E) (Fig. 14). The standard deviations were again high, but the means differed to such an extent that there was a trend in favour of xylitol. The enzyme activities were expressed in U/litre and the results were thus similar to those obtained with monkey parotid saliva (Bird *et al.*, 1977). The findings with human saliva may not have any clinical significance.

SALIVARY CARBONIC ANHYDRASE AND THE CONSUMPTION OF XYLITOL

Carbonic anhydrase (carbonate hydrolase, EC 4.2.1.1) which catalyses the reaction

$$H_2CO_3 = CO_2 + H_2O$$

became the object of attention in this laboratory after it was found that the chewing of xylitol-sweetened chewing gum resulted in a higher concentration of HCO_3^- ions in human whole saliva than the use of chewing gum sweetened with other sugars (Fig. 15). A recent study on 13 human subjects showed that 15-min chewing of xylitol-sweetened fruit pastilles did not significantly increase the activity of parotid saliva carbonic anhydrase when compared with the effect caused by a corresponding sucrose product, when the results were expressed in units/ml (Fig. 16) (Mäkinen *et al.*, 1977E). The bicarbonate system comprises approximately 85% of the total buffering capacity of saliva. The HCO_3^- ions thus represent an

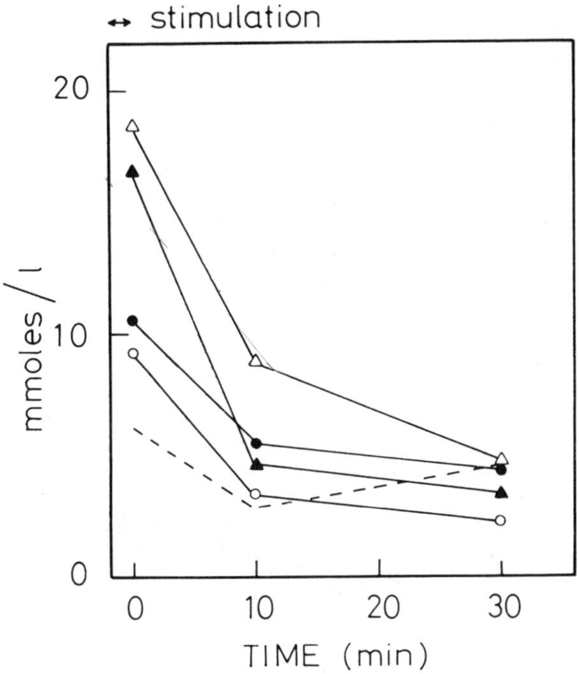

FIG. 15. Salivary HCO_3^- ion concentration as a result of 2-min stimulation with sweetened and unsweetened stimulators. The arithmetic means of values from 10 subjects are given. The stimulation was accomplished with chewing gums sweetened by the sugars indicated. ●, chewing gum base; ----, paraffin; △, xylitol; ○, sucrose; ▲, sorbitol. (Söderling *et al.*, 1975.)

important entity in saliva and it is possible that part of the increase in the concentration of HCO_3^- ions mentioned above resulted from a flux of ions through the oral mucosa as a result of the osmotic properties of xylitol (Mäkinen, 1978).

CHANGES IN THE pH VALUES OF SALIVA IN RELATION TO USE OF SUGARS

When human subjects rinse their mouths with a fermentable sugar solution, the pH value of dental plaque usually decreases steeply. The minimum pH value normally appears 10 min later. The pH of

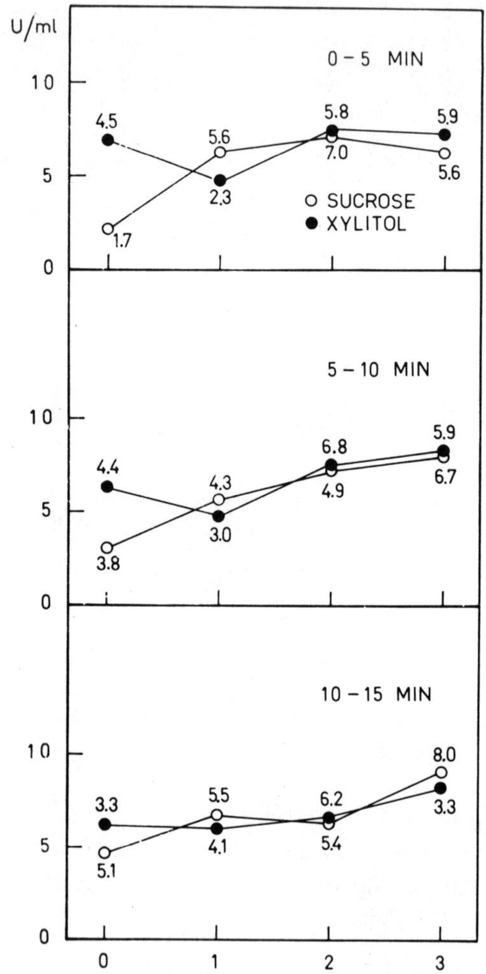

FIG. 16. Effect of chewing of xylitol- or fructose-containing fruit pastilles on the activity of carbonic anhydrase of human parotid saliva. Details are as for Fig. 3. (Mäkinen *et al.*, 1977E.)

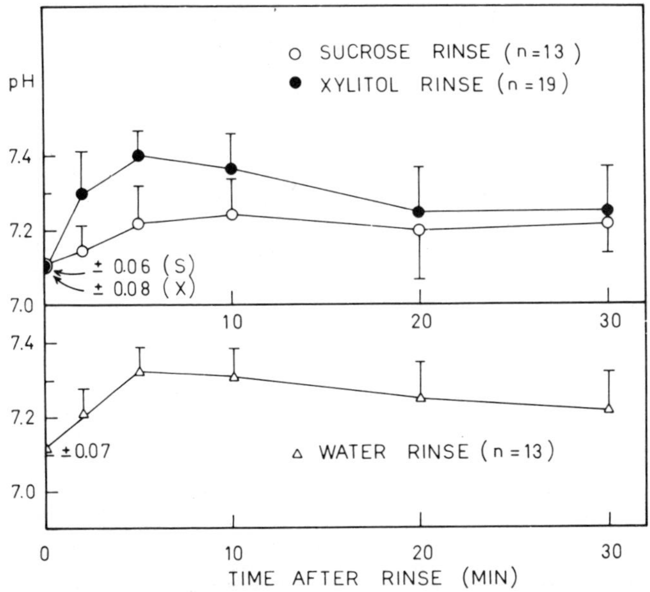

FIG. 17. Effect of xylitol and sucrose mouth rinses on the pH values of centrifuged human whole saliva. The details are described in Fig. 7.

plaque then returns close to the original value during a period of 45–60 min provided that no additional sugar is taken in. The study of plaque pH values in this way gives the so-called Stephan's curve which is classic in oral biology. The more frequently this type of pH minimum occurs and the lower the value attained each time, the more likely it is that demineralisation of the enamel structure at particular tooth sites will take place.

Saliva attempts to correct such pH drops. So, for example, if the pH value of whole saliva is determined instead of that of plaque, one usually observes an initial increase in pH even in case of readily fermentable sugars. All sugars normally increase the flow rate of saliva and produce a transitory pH rise in it. With fermentable sugars the effect is, however, usually counteracted by the acids formed by the bacteria as a result of glycolysis and other fermentations. The dietary sugars differ, however, to a noticeable extent in their ability to elicit the above-mentioned initial pH rise in

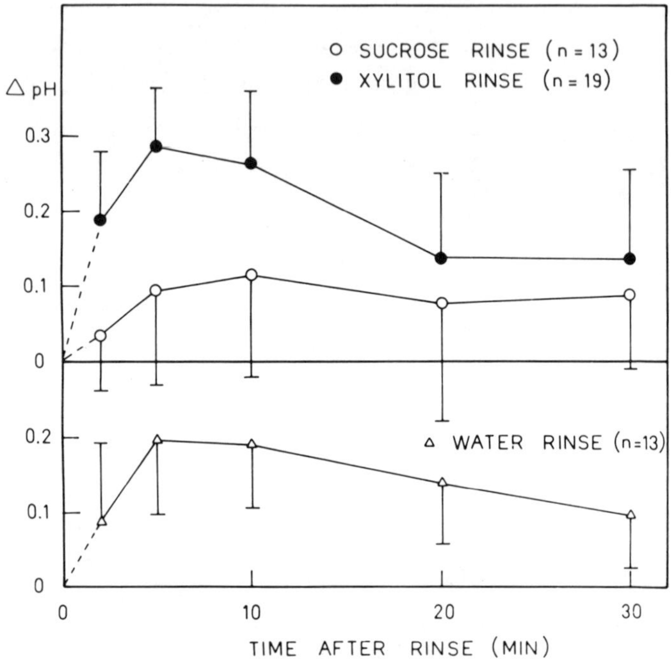

FIG. 18. Data of Fig. 17 plotted as differences of pH values compared to the initial value determined before rinse.

saliva. Even more pronounced differences in pH values can be encountered at later stages after the exposure of the oral tissues to different sugars. Non-fermentable sugars maintain a high alkaline pH value in the oral fluid considerably longer than the fermentable ones. This can be shown for sucrose and xylitol in Fig. 17. The rinse with a 10% xylitol solution (15 ml) for 2 min with simultaneous paraffin chewing produced a significantly higher maximum pH value in oral fluid than sucrose or water. If the antilog_{10} of the maximum pH values recorded for sucrose, xylitol and water are calculated, the following hydrogen ion concentrations are obtained: S, $5{\cdot}75 \times 10^{-8}$ M; X, $3{\cdot}98 \times 10^{-8}$ M; H_2O, $4{\cdot}79 \times 10^{-8}$ M. Consequently, the sucrose rinse caused a maximum hydrogen ion concentration which was 30% higher than that caused by xylitol rinse. The difference between sucrose and xylitol can also be seen in

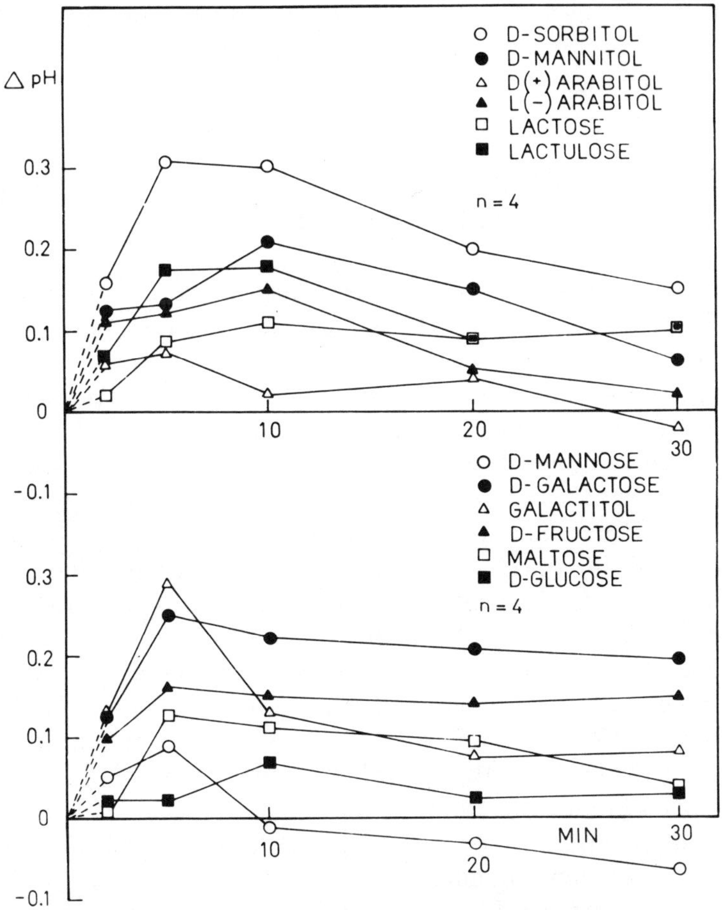

FIG. 19. Effect of mouth rinses with various sugar solutions on the pH values of human whole saliva. Data from a pilot study ($n = 4$) performed as described in Fig. 7.

Fig. 18 which gives the differences of the pH values, calculated with the aid of the starting values (before rinse). The difference between water and xylitol solution was also considerable. It is important that the pH-raising effect of xylitol is rather quickly manifested in saliva after rinsing.

Figure 19 shows the results from a pilot study. Rinsing the mouth with sorbitol produced higher pH values than rinsing with several other sugar solutions. Readily fermentable hexoses produced a slight pH rise only. The details of this study are as for those described in Fig. 7.

OTHER CHEMICAL CHANGES IN SALIVA IN RELATION TO CONSUMPTION OF XYLITOL

A number of biochemical parameters of whole saliva (oral fluid) were studied in connection with the Turku sugar studies (Mäkinen and Scheinin, 1975A). Because whole saliva was used, many of the findings made concerned enzymes and compounds, the levels of which in saliva depended on the effect of xylitol on oral microorganisms. Hence only indirect or scant information about the chemical responses of true saliva was obtained. However, in some cases, as in case of lactoperoxidase, true salivary responses could be revealed. Both pure secretions and whole saliva have their advantages as test material. To study the latter is also rational, as it is the whole saliva (oral fluid) which ultimately bathes the oral tissues, but one should also elucidate the contribution of the separate salivary glands to oral biology. The following findings add new information to the previous ideas about the effects of xylitol saliva.

Rinsing the mouth with 10 % solutions of sucrose or xylitol (15 ml) did not significantly affect the concentration of inorganic phosphorus (Fig. 20), calcium (Fig. 21) and proteins (Fig. 10) in whole saliva when compared to rinsing with an equal volume of water. The experimental details of this study were described earlier (Fig. 7). In another study, human subjects rinsed their mouths with a 10 % solution (15 ml) of xylitol or sucrose at one-hour intervals over a period of 6 h. Each rinse lasted for 5 min and the subsequent whole saliva samples (4·0 ml) were collected each hour by paraffin stimulation. The flow rate was recorded. No significant differences between the two sugars were found in the parameters studied (sulphate and sialic acid, both determined after acid hydrolysis of saliva, and α-amylase). The concentration of proteins in whole saliva during rinsing with xylitol remained at a somewhat lower level than during rinsing with a sucrose solution. This may be due to the fact

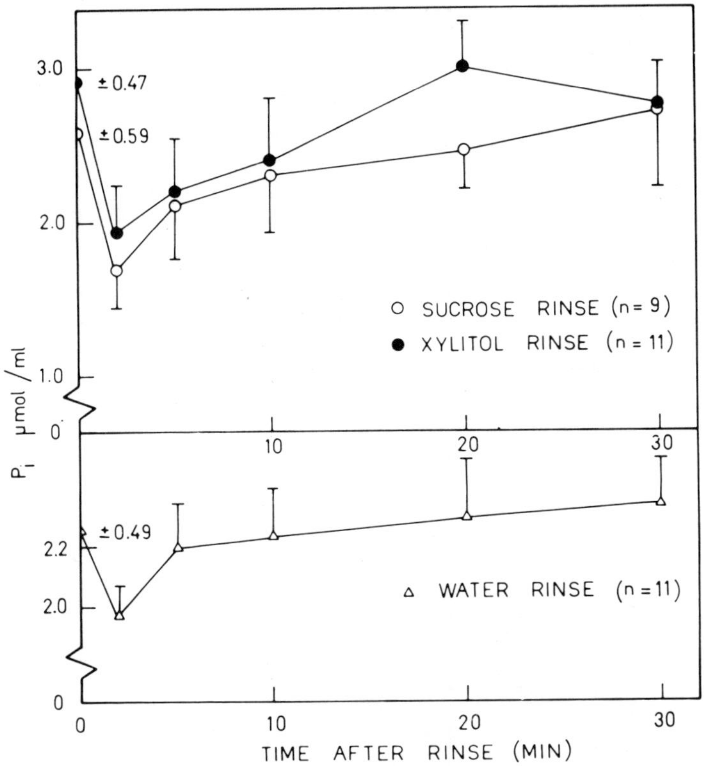

FIG. 20. The concentration of inorganic phosphorus in centrifuged (23 500 × *g*) whole saliva samples after rinsing the mouth with 10% solutions of xylitol or sucrose. The details of the study are described in Fig. 7.

that xylitol rinses are able to reduce the growth of dental plaque to a certain extent. Consequently, although the pH values of whole saliva were readily and selectively affected by rinsing with sugar solutions, other chemical parameters remained almost unaffected. Single mouth rinses with 10% sugar solutions elicit selective effects on a very restricted number of parameters. Long-term administration of higher amounts of sugars may cause additional effects.

A particular pH-raising factor in saliva has been described (Kleinberg, Kanapka and Craw, 1976). This factor is a small peptide, sialin, having the following amino acid composition: glycine–glycine–lysine–arginine. Sialin may account for most of the

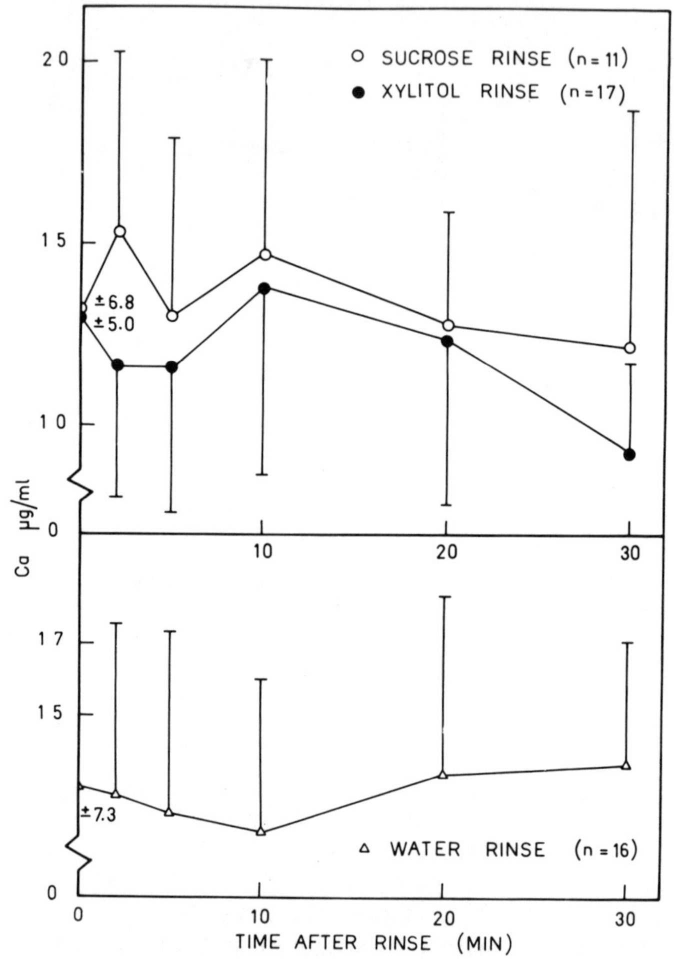

FIG. 21. The concentration of calcium in centrifuged (23 500 × *g*) whole saliva samples after rinsing the mouth with 10 % solutions of xylitol or sucrose. The details of the study are described in Fig. 7.

pH-raising activity of salivary supernatant fluid. It was not possible to isolate this factor from the whole saliva samples of the subjects in the Turku sugar studies. It is nevertheless interesting that although the consumption of a xylitol diet increased the concentration of all amino acids in whole saliva, it was the basic amino acids (arginine, lysine and histidine) and glycine which were strongly increased. This suggests that the xylitol consumption was associated with the formation of a high number of basic equivalents in the oral cavity. This may be important when considering the neutralisation of acids formed by glycolysis in plaque (Mäkinen, Lönnberg and Scheinin, 1975). The Turku sugar studies did not show, however, any direct relationship between xylitol consumption and secretion of the salivary pH-raising factor, but the results suggested that detailed experiments along this line should be carried out.

Arginine solutions give a red colour upon treatment with α-naphthol and sodium hypochlorite. This reaction, usually termed the Sakaguchi reaction, is due to the presence of the guanidino group in arginine. Because arginine is the only protein amino acid known to contain this group, the Sakaguchi reaction has been used to estimate the amount of arginine present in protein hydrolysates. Arginine covalently bound to native peptide chains also reacts, however, in the presence of the reagents mentioned, although not to such an extent as arginine in protein hydrolysates or in denatured proteins. Furthermore, although the reaction is rather specific, certain interference is caused by molecules closely resembling the guanidino group. The estimation of Sakaguchi-positive compounds in saliva provides, however, valuable information. This type of experiment was carried out on human subjects in this laboratory. Parotid saliva samples were collected after initial stimulation with fruit pastilles sweetened with either sucrose or xylitol. The samples were collected during a period of 3 h after stimulation. The numbers of samples collected after sucrose and xylitol pastilles were 140 and 148, respectively. The Sakaguchi reaction produced the following results (expressed as μg arginine/ml parotid saliva): xylitol, 1217 ± 426; sucrose, 1078 ± 450. The data indicated no significant differences between the effects caused by sucrose and xylitol, although there was an increasing trend in favour of xylitol. Chewing of xylitol-containing fruit pastilles during a 15-min period did not thus yield more than an indicative change in Sakaguchi-positive material in saliva. Details of this study have been reported separately (Mäkinen *et al.*, 1977E).

The arginine values shown above need the following consideration.

The analysis determined all arginine-like compounds present in the samples, including both free and protein-bound arginine. Consequently, the values are much higher than those given for free salivary arginine. The latter values are of the order of 18–32 μmol/l (supernatant fluid of whole saliva) (Mäkinen, Lönnberg and Scheinin, 1975), or 0–13 μmol/l (supernatant fluid of freshly secreted saliva) (Kleinberg, Kanapka and Craw, 1976). The concentration of free arginine in saliva is thus low, but arginine-containing peptides and proteins are present in ample amounts.

The contribution of enteral xylitol to oral biology has recently been touched upon in greater detail (Mäkinen, 1978).

LACHRYMAL FLUID

Monkey lachrymal fluid has a considerable lactoperoxidase activity (Mäkinen *et al.*, 1977A), but nothing is known about the effects of xylitol diet on the properties of this fluid as compared with those caused by a sucrose diet. Preliminary information suggests that gastric intubation of either xylitol or sorbitol does not influence the peroxidase activity of monkey tears. A 12-week feeding study in cows was commenced in the autumn of 1976. The animals were given 0·5 kg a day of a mixture of sugar alcohols. The mixture contained chiefly sorbitol, mannitol, arabitol and xylitol with smaller amounts of other polyols. Samples of lachrymal fluid were collected biweekly and analysed for peroxidase, aminopeptidase (with *N*-L-arginyl-2-naphthylamine as substrate) and protein. The information obtainable at this stage suggests that no differences between the polyol, molasses and control groups were found (Mäkinen *et al.*, 1977D).

The study of lachrymal fluid and other exocrine secretions has been a logical consequence of the findings made with saliva. It was thought that if a particular type of exocrine gland responds to polyol consumption, other functionally or structurally similar glands should also be studied, no matter how unlikely a relationship between the gland involved and consumption of xylitol would be.

MILK

Evidence indicates that milk proteins are synthesised by mammary tissue from amino acids furnished by the blood. The above-mentioned feeding study on cows was thus extended to comprise an investigation of the levels of milk proteins, electrolytes and certain other compounds. The present information suggests that the consumption of the polyol diet was associated with a higher and more constant lactoperoxidase activity than was determined in the molasses and the control groups (Mäkinen *et al.*, 1977D). The electrolytes (Na^{+}, K^{+}, Ca^{2+}, Mg^{2+}, Fe^{2+}) remained unchanged.

Because of the complex composition of the polyol mixture, it was not possible to conclude which sugar alcohol had produced the strongest effect, or whether all polyols were responsible for the higher enzyme level.

The milk studies evoke the following interesting possibility in animal feeding. If it were conceivable that sugar alcohols like xylitol have an advantageous effect on milk, one would achieve favourable changes in the growth and physiological properties of the progeny. Such an idea should naturally be subjected to thorough investigation. The somewhat higher milk lactoperoxidase activity of polyol-fed cows may affect the preservation qualities of milk before processing. A pilot study on human milk has been started in this laboratory.

CONCLUSION

The biochemical changes in the properties of exocrine secretions described in this article should be understood as normal and physiological responses to changes of the composition of diet. In organs whose secretions contain more than one enzyme, e.g. the salivary glands, it is possible that individual cells synthesise more than one enzyme. However, on repeated stimulation of the secretory activity of the pancreas, the relative amounts of the most important enzymes seem to remain constant in the pancreatic juice. The present and previous results suggest that prolonged stimulation of the salivary glands and pancreas does not exhaust the glandular ability to liberate a secretion of almost constant electrolyte content. Prolonged stimulation may, however, exhaust the supply of enzymes and mucins. This is reflected in the initial decrease of the levels of lactoperoxidase, for example, in saliva after stimulation with xylitol chewing gum. When enough time for the protein synthesis has been given (2–3 h), stimulation with xylitol may even exceed the lactoperoxidase levels compared with stimulation by sucrose. The present results suggest that the consumption of xylitol in a proper predetermined way is associated with increased levels of lactoperoxidase in saliva. This effect may have advantageous effects in oral health.

The mechanism of the xylitol effect in caries prevention can be explained for the most part. As described elsewhere in more detail (Mäkinen, 1978), the decisive properties of xylitol in this sense are its essential non-fermentability in human dental plaque and its selective physiological influence on the minor and major salivary glands. Many other sugar alcohols may share even both of the above properties with xylitol, but the high sweetness, pleasant organoleptic qualities and other biochemical aspects of xylitol seem to place it first among the calorific sweeteners currently available. Comparisons with sorbitol and xylitol have been presented earlier (Mäkinen, 1976 and 1978).

The study of the effects of peroral administration of xylitol on the function of other exocrine glands is in the initial phase. The research should be extended to comprise the whole exocrine system, as xylitol and certain other sugar alcohols may exert selective effects to a greater extent than previously realised.

The effects of perorally administered xylitol on the major exocrine glands is restricted by the slow absorption rate of xylitol and its rapid turnover to glucose in the liver. Direct information about the effects could be obtained in connection with continuous intravenous administration of xylitol.

REFERENCES

Afonsky, D. (1961) In: *Saliva and its Relation to Oral Health*, p, 154. Alabama: University of Alabama Press.

Basu, T. K., Dickerson, J. W. T. & Parke, D. V. (1975) Effect of dietary substitution of sucrose and its constituent monosaccharides on the activity of aromatic hydroxylase and the level of cytochrome P-450 in hepatic microsomes of growing rats. *Nutr. Metab.*, **18,** 302–9.

Bird, J. L., Baum, B. J., Mäkinen, K. K., Bowen, W. H. & Longton, R. W. (1977) Xylitol associated changes in amylase and protein content of monkey parotid saliva. *J. Nutr.*, **107,** 1763–7.

Chandra, T., Das, R. & Datta, A. G. (1977) Role of thyroid gland on the peroxidase and iodinating enzymes of submaxillary gland. *Eur. J. Biochem.*, **72,** 259–63.

Harper, L. R., Poole, A. E. & Wolf, S. I. (1977) Xylitol stimulation of lactoperoxidase in human parotid saliva. *55th Session of the Internat. Assoc. for Dental Research*, Copenhagen, Abstract No. 78.

Herman, R. H. (1974) In: *Sugars in Nutrition*, eds. H. L. Sipple & K. W. McNutt, pp. 145–72. New York: Academic Press.

Hoogendoorn, H., Piessens, J. P., Scholtes, W. & Stoddard, L. A. (1977) Hypothiocyanate ion; the inhibitor formed by the system lactoperoxidase–thiocyanate–hydrogen peroxide. I. Identification of the inhibiting compound. *Caries Res.*, **11,** 77–84.

Kleinberg, I., Kanapka, J. A. & Craw (1976) Effect of saliva and salivary factors on the metabolism of the mixed oral flora. In: *Proceedings on Microbial Aspects of Dental Caries*, eds. H. M. Stiles, W. J. Loesche & T. C. O'Brien. Sp. Suppl. *Microbiology Abstracts*, Vol. **II,** 433–64.

Mäkinen, K. K. (1976) Microbial growth and metabolism in plaque in the presence of sugar alcohols. In: *Proceedings on Microbial Aspects of Dental Caries*, eds. H. M. Stiles, W. J. Loesche & T. C. O'Brien. Sp. Suppl. *Microbiology Abstracts*, Vol. **II,** 521–38.

Mäkinen, K. K. (1978) Biochemical principles of the use of xylitol in medicine and nutrition with special consideration of dental aspects. *Experientia*, Suppl. 30, in press.

Mäkinen, K. K. & Scheinin, A. (1975A) Turku sugar studies. VII. Principal biochemical findings on whole saliva and plaque. *Acta Odont. Scand.*, **33,** Suppl. 70, 129–71.

Mäkinen, K. K. & Scheinin, A. (1975B) Turku sugar studies. XIII. Effect of the diet on certain clinico–chemical values of serum. *Acta Odont. Scand.*, **33,** Suppl. 70, 265–76.

Mäkinen, K. K. & Scheinin, A. (1975C) Turku sugar studies. XIX. Salivary peroxidase and invertase-like activity in relation to 1 year use of sucrose and xylitol chewing gum. *Acta Odont. Scand.*, **33,** Suppl. 70, 317–20.

Mäkinen, K. K., Bowen, W. H., Dalgard, D. & Fitzgerald, G. (1977A) Effect of peroral administration of xylitol on exocrine secretions of monkeys. *J. Nutr.*, in press.

Mäkinen, K. K., Läikkö, I., Rekola, M. & Scheinin, A. (1977B) Biochemical changes in plaque and whole saliva in relation to one month consumption of xylitol and sorbitol chewing gum. To be published.

Mäkinen, K. K., Lönnberg, P. & Scheinin, A. (1975) Turku sugar studies. XIV. Amino acid analysis of saliva. *Acta Odont. Scand.*, **33,** Suppl. 70, 277–86.

Mäkinen, K. K., Mäkinen, P.-L., Söderling, E. & Tenovuo, J. (1977C) Effect of peroral administration of xylitol on the peroxidative and dehydrogenation potential of human blood cells. To be published.

Mäkinen, K. K., Tenovuo, J. & Scheinin, A. (1975) Turku sugar studies. XII. The effect of the diet on oral peroxidase, redox potential and the concentration of ionized fluorine, iodine and thiocyanate. *Acta Odont. Scand.*, **33,** Suppl. 70, 247–63.

Mäkinen, K. K., Tenovuo, J. & Scheinin, A. (1976) Xylitol induced increase of lactoperoxidase activity. *J. Dent. Res.*, **55,** 652–60.

Mäkinen, K. K., Tuori, M., Poutiainen, E. & Hämäläinen, M. (1977D) The effect of the consumption of a polyol mixture on serum, milk, whole saliva and lacrimal fluid in the cow. To be published.

Mäkinen, K. K., Virtanen, K., Mäkinen, P.-L. & Neva, M. (1977E) Biochemical changes in human parotid saliva after single dosage of xylitol. To be published.

Meyer, K. H., Fischer, E. H., Bernfeld, P. & Duckert, F. (1948) Purification and crystallization of human pancreatic amylase. *Arch. Biochem.*, **18,** 203–5.

Morrison, M. & Steele, W. F. (1968) In: *Biology of the Mouth*, pp. 89–110. Washington: American Association for the Advancement of Science.

Söderling, E., Rekola, M., Mäkinen, K. K. & Scheinin, A. (1975) Turku sugar studies. XXI. Xylitol, sorbitol, fructose and sucrose induced physico-chemical changes in saliva. *Acta Odont. Scand.*, **33,** Suppl. 70, 337–43.

Discussion: Biochemical Aspects

Chairman—PROFESSOR A. E. BENDER

A. E. Bender (*Queen Elizabeth College, London University, UK*): In our own work we established long ago that the liver is enlarged on sucrose feeding due to the fructose moiety. New cells are stimulated and the kidney is enlarged. The histological changes in the kidney are pathological and quite distinct from any enzymic changes. They are major changes in the tissue of the animal. Have you examined the experimental animals after feeding xylitol?

H. Förster (*Klinikum der Johann Wolfgang Goethe University, Frankfurt-am-Main, W. Germany*): We have examined them after xylitol feeding and there are no gross pathological changes, even after long lasting infusion of xylitol in high doses. Some enzymic differences are found but this is not true for oral feeding even if 20% of the animal's total feed is xylitol. I think the changes which occur in sucrose feeding only happen if it is given in amounts of 50–80% of the feed. We found these differences then but did not find them when only 20% of the feed was sucrose or fructose.

K. K. Mäkinen (*Institute of Dentistry, Turku University, Finland*): When we were performing the two-year Turku sugar study, we analysed liver function in several accepted ways, for example, by determining the transaminase activity and so on. All of the clinical chemical values determined during the whole two-year study were quite the same in all these three sugar groups. Thus, no matter what the subsequent use of sucrose, fructose and xylitol, the values were the same.

V. Marks (University of Surrey. Guildford, UK): What is the effect of xylitol on alcohol metabolism? Does it slow it down as one would expect it might?

Förster: We have performed studies with human volunteers who were given ethanol and xylitol and there was no slowing down of ethanol metabolism. There are, however, great differences between the metabolism of ethanol when combined with xylitol or sorbitol, even though ethanol has a strong influence on galactose metabolism. Galactose is a sugar which is metabolised by the liver and the epimerase which is the important enzyme for this is NAD dependent.

If ethanol is metabolised, then NADH is produced and the epimerase is not functionally active and during ethanol administration the galactose metabolism is inhibited to a very high extent. We have done the same experiment with xylitol and sorbitol, not in equal doses but doses where equal NAD and NADH production are found and there was no effect by xylitol or by sorbitol on galactose metabolism. This means that different ways of oxidation in the liver parenchyme are found by the different alcohols and ethanol and the polyalcohols are not comparable in metabolic effects.

Comment: I noted from your paper that on a sucrose diet, saliva had a lower pH. I always thought that the problem with sucrose was that it reduced the pH.

Mäkinen: Well, perhaps I did not explain it thoroughly. We examined the effects of sugars on the pH values of plaque and saliva. If we determine the pH values in dental plaque we see almost immediate reduction when the sucrose is consumed but it is not wrong that sweeteners can cause a temporary increase in the pH values of saliva which are selective. Certain sugars are able to produce higher pH values than other sugars but it is not a discrepancy at all so we have to differentiate between pH values of plaque and saliva. These are quite different biological objects to study but the interesting thing is that among the sugars there are certain specific sugars like sorbitol and xylitol which are clearly able to increase the pH values more, but with fermentable sugars like sucrose, glucose and fructose the pH values very rapidly go down and reach the initial values again, but with polyalcohols the level of pH is maintained at the higher stage during a prolonged period. Thus many sweeteners can cause an initial increase in pH. It has nothing to do with a possible non-cariogenic nature of sucrose, the pH values are fairly small and they rapidly come back again to the initial level.

C. Stephens (*Nicholas Labs., Bath Road, Slough, UK*): May I follow that up? We have this problem of salivary pH being fluid-flow dependent. If we consider parotid flow as the main component of the whole salivary collection during stimulation, then what we could be getting here is a slowly dependent effect on availability of the stimulus, i.e. each sugar could be stimulating the parotid separately and, therefore, altering your total pH components here. This is due to increased parotid flow which will automatically increase the pH.

Mäkinen: That is certainly true. In the monkey study we used pilocarpine to stimulate the salivary flow which also has its own effects. Pilocarpine was used in both sucrose and xylitol groups but we also used paraffin stimulation because otherwise it is impossible to collect enough saliva; so a partial answer to this question must be that as we had to use paraffin stimulation this too might have increased the pH values temporarily.

Question: Could you tell me if there is any change in the levels of iditol dehydrogenase with continued administration of xylitol?

K. H. Bässler (*Johannes Gutenberg University, Mainz, W. Germany*): This has been measured only in animals; in rats there is an adaptation of the iditol dehydrogenase. It increases after continuous administration of xylitol and, maybe because of this adaptation of the absorption from the gut, it causes a more favourable concentration gradient from the intestinal mucosa and goes out through the blood.

G. L. S. Pawan (Middlesex Hospital Medical School, London, UK): In your long-term studies of man, did you notice any changes in the cholesterol and triglyceride levels? You did mention that there were normalising effects on the free fatty acids as well as an antiketogenic effect.

Förster: During oral administration there are no effects in the animal. There are effects in the human being if an intravenous administration is used in high dosage. An increase in triglyceride level is found if any kind of sugar is given. There are differences between sugars. If glucose is given intravenously, there is still a blood level 20–28 h after the infusion but with xylitol after an infusion, 8–10 h. This is the only difference and the order of magnitude is the same in both cases.

In animal-feeding experiments we did not find any increase in triglyceride levels and, in fact, found a decrease if up to 20% of the feed was given in the form of xylitol.

If high doses of xylitol were given intravenously, lasting for three days, a small increase in triglyceride levels was found as well as a small accumulation of triglycerides in the liver. These, I must mention again, are only found if very high doses are given intravenously.

Human volunteers have been given up to 600 g per day for two days via the intravenous route and animals can be given even higher doses.

Regarding cholesterol, there were no increases in the levels and we have found no signs of increase after xylitol or sorbitol. However, with oral feeding of xylitol there is a small increase compared with sucrose.

Mäkinen: We also found in our two-year Turku study that there was no increase in triglyceride level in human volunteers either.

Pawan: What is the renal threshold of xylitol in man? Does it spill over into the urine when you give it intravenously in any considerable amounts so that the blood level is about 100 mg%.

Bässler: There are actually no renal thresholds of xylitol. As soon as you have a xylitol level in the blood you find renal excretion but this is not pertinent to oral application because if you give xylitol by mouth practically the whole amount is taken up by the liver and it does not go through to the general circulation. This is different if you give it intravenously.

G. Frostell (*Lund University, Malmö, Sweden*): Are you aware of any studies on possible gross changes or microscopic changes of the intestinal mucosa as caused by high oral doses of sugars or sugar alcohols?

Mäkinen: Well, I may have a partial answer based on a literature review that I have been carrying out recently. According to a survey of 600 papers, I have not found any study which showed any detrimental or negative effects on the intestinal mucosa after loading the subject with xylitol. I think that the mucosa is adapted to enormous changes whatever we may have in our diet.

H. Hummler (*Hoffmann-la-Roche, Basle, Switzerland*): We have current long-term toxicological studies using rats, mice and dogs and after duration of 78 weeks we performed interim sacrifice of the animals and could not find any lesions of the intestinal mucosa. I can add something on the previous question in that we did not find after the 78 weeks of the experiment any increase in triglycerides or the cholesterol plasma levels.

D. A. M. Geddes (*Glasgow Dental Hospital & School, Glasgow, UK*): Are there any medical conditions where the oral administration of xylitol might be contraindicated?

Förster: In my opinion, there are no contraindications for the oral administration of xylitol. However, there is a limitation of the dose. We have performed experiments with children and found that it is possible to give children 30 g or more per day but that a single dose should not exceed 10–15 g. Until now, we have not made any experiments with confectionery. I think that sweets can be tolerated in up to 20 g as a single dose. The only contraindication is the limitation of the dose because if the dose is too high then diarrhoea might occur. This diarrhoea might also limit the intake of xylitol in the form of candy.

Mäkinen: It is necessary to emphasise one thing. In order to achieve the beneficial dental effects, the dosages required are small and thus the possibility of diarrhoea due to over dosage should be controlled. If we used chewing gum or lozenges purely for dental reasons there would be no reason for diarrhoea.

Question: Have any experiments been carried out using xylitol-based compounds rather than xylitol as a pure substance? And if they have, are the results obtained similar to that using xylitol on its own?

Mäkinen: To my knowledge, there have been synthetic compounds derived from xylitol like xylitol pentanitrate. And, although there are many compounds known, they are not being tested and there would be no reason to do so. Xylitol has already been shown to be a natural compound and we already know the physiological metabolic routes. We have no reason to test synthetic compounds.

I do know that certain xylitol compounds can alleviate the effects of catalin poisoning and some xylitol compounds can be effective against thyroid infections and so on, but this is all that I know about them.

Question: You have shown that after consumption of xylitol, lactoperoxidase levels and amylase levels are enhanced. What is the beneficial effect of the amylase increase on dental caries?

Mäkinen: This is an important question. Amylase is a digestive enzyme which partly breaks down carbohydrates and, therefore, if there was very high amylase activity, this could produce unnecessary amounts of fermentable units in the oral cavity. I do not have any proof of this possibility to show that increased amylase levels would be in any way harmful as it is after all a physiological enzyme. I must admit that this should be studied. However, it is my firm belief that any increase of alpha-amylase activity would still be within the normal physiological range and several other dietary ingredients as well as xylitol are able to increase this level.

Question: Have there been any studies on the possible bacterial breakdown of xylitol in the colon?

Mäkinen: I only know of one study related to the effect of xylitol on colonic bacteria. This has shown that it is more or less non-fermentable by the microorganisms commonly found in man. They simply do not attack xylitol, and, if they do, it is at a very low rate and to a very small extent.

CHAPTER 7

Cariogenic Bacteria

F. GEHRING
(*Bayerische Julius-Maximillians-University, Würzburg, W. Germany*)

NATURE OF DENTAL CARIES—GENERAL CONSIDERATIONS

In the civilised countries of our earth, dental caries takes first place in the statistics of human illnesses. The problems related to this illness are reviewed extensively in textbooks (Burnett and Scherp, 1968; Nolte, 1968 and 1973; Skinner and Carr, 1974; Stiles, Lösche and O'Brien, 1976) and articles (Ellen, 1976; Gibbons and van Houte, 1975). The origin of dental caries depends on the consumption of carbohydrates, on the presence of microorganisms in the oral cavity, and their contact with the teeth surfaces. Development of caries is solely possible by participation of *all* these three factors: teeth of the host; carbohydrates in the diet; and microorganisms.

MICROBIOLOGY OF THE ORAL CAVITY

An immense number of microorganisms can be found in every oral cavity, which they reach via food and inhalation. Humidity and warmth in the oral cavity are excellent conditions for the growth of microorganisms, especially when larger amounts of food particles are at their disposal as a consequence of faulty oral hygiene. As humans and microorganisms have nearly the same dietary requirements, the most nutritious foods, meat, milk, potatoes, sugar, etc., are also the most valuable nutritive substances for microorganisms. Due to their particular characteristics they have several advantages over the human host in the competition for the same nutrients. For instance, microorganisms are able to take up nutritive substances across the entire cell surface. In 24 h they can metabolise quantities of nutrients amounting to several times their own weight. The metabolism of microorganisms can be 200–2000 times as efficient as the comparable metabolism of the stomach or liver of humans. This enormous metabolic efficiency is mainly possible because in microorganisms the ratio of surface to volume or weight is relatively large. In addition, microorganisms are very small and have a huge multiplying capacity. Under favourable growth conditions the number of cells doubles every 20–30 min. It is not surprising that several hundreds of millions of organisms can be found in 1 ml saliva and that this number is as much as a hundred to a thousand times higher in the corresponding amount (1 g wet weight) of the film—known as plaque—found on the teeth. The total amount of microorganisms present in the oral cavity of humans can sum up to 50×10^{12}. It should be remembered that not all microorganisms present in the oral cavity can be cultivated *in vitro* but, as can be determined by microscopy, the actual number of organisms in different regions is essentially higher.

About fifty known species of microorganisms have been isolated from the oral cavity of humans so far. The different species do not occur with the same frequency. According to the location some species dominate, while others are scarce. But in general the qualitative and quantitative microbial composition of the oral cavity is more or less balanced, although local and temporary fluctuations are possible.

THE MICROBIOLOGY OF DENTAL CARIES

Seldom does a scientific statement retain its validity as long as the chemico-parasitic theory of caries set up by Miller at the end of the

19th Century (Miller, 1890). This theory states that caries is the result of a demineralisation process under the influence of acids and that these acids originate from the microbial degradation of carbohydrates. The theory is still valid in principle today. Acids thus have a key position in this process and indeed most oral microorganisms are capable of producing acids by metabolising a number of carbohydrates in order to maintain their own metabolism. These acids are formed primarily in the inner layers of the dental plaque, in which anaerobic conditions prevail. For this reason dental plaque is considered to be the area of caries initiation and hence is of central significance to caries aetiology and is subject to intensive studies (McHugh, 1970). In microbiological terms, plaque consists of a mixed bacterial flora living in a complex nutrient medium (food residues). At the beginning of plaque formation Gram-positive cocci and bacilli dominate, later also Gram-negative cocci and bacilli, and eventually filamentous organisms such as actinomyces, leptothrix, as well as vibrio and spirochaetes.

Cariogenic Microorganisms

Microbiological caries research pays special attention to the streptococci which predominate just as much at the beginning of plaque development as in older plaque. Certain species of this group have been studied more and more frequently since the beginning of the sixties. The increasing application of gnotobiotic techniques in animal experiments concerned with caries, which have tested the cariogenic properties of a great number of microorganisms by monoassociation in germ-free rats, has led to this development. It turned out that certain streptococci with particular characteristics had such cariogenic properties. Carlsson (1967) re-introduced the species name *Streptococcus mutans* for these organisms. This name had been proposed by Clarke (1924) for a number of streptococci which he had isolated from carious lesions of teeth and which could not be classified. Since about 1967 several synonyms for this organism have been used by the authors of literature on this subject. A survey of these synonyms is given in Fig. 1. Today seven serotypes and several biotypes and genotypes of *S. mutans* are known. But the latest edition of *Bergey's Manual of Determinative Bacteriology*

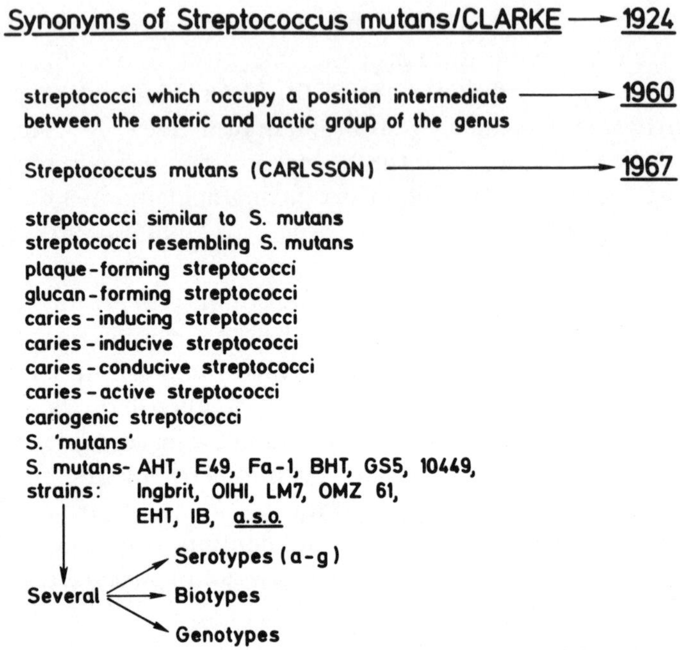

FIG. 1.

(Bergey, 1974) does not describe this species, which seems to suggest that its taxonomic position has remained a problem. Nevertheless, the name '*mutans*' is generally used and in recent years the knowledge of its characteristics has increased significantly. The cariogenic properties of other microorganisms apart from streptococcal species have been tested by gnotobiotic techniques. Some of the results of such investigations are reviewed in Fig. 2 (Nolte, 1973). Accordingly, cariogenic properties could also be determined for other streptococci, some lactobacilli and actinomycetes species. On the other hand, several species proved to be non-cariogenic, among them again streptocci and lactobacilli species. This shows that differences even between the species of one genus do exist.

Acid Production

For caries aetiology it is of great importance to know the

STREPTOCOCCUS	mutans, numerous strains	Cariogenic
	faecalis	Cariogenic
	liquefaciens	Cariogenic
	mitis	Cariogenic
	salivarius-like	Cariogenic
	sanguis	Cariogenic
Lactobacillus	acidophilus	Cariogenic
	casei	Cariogenic
	unknown species	Cariogenic
Actinomyces	viscosus	Cariogenic
	naeslundii	Cariogenic
Lactobacillus	species	Noncariogenic
Streptococcus	species	Noncariogenic
Clostridium	tertium	Noncariogenic
Staphylococcus	species	Noncariogenic
Proteus	vulgaris	Noncariogenic
Aerobacter	aerogenes	Noncariogenic
Pseudomonas	aeruginosa	Noncariogenic
Leuconostoc	dextranicum	Noncariogenic
	mesenteroides	Noncariogenic

FIG. 2. Cariogenic and non-cariogenic microorganisms. (Tested on rats, hamsters, or both.)

characteristics that enable certain bacteria to develop cariogenic properties. Above all, there is the effect of the ability to produce larger amounts of acids in a short time, especially from low molecular weight carbohydrates. In this connection a characteristic of *S. mutans* (Fig. 3) should be mentioned. *S. mutans* can still produce acids even with 40% sucrose in the culture medium. *Streptococcus sanguis* and *Streptococcus salivarius* do not show this effect (Gehring, 1972, 1977). The well-known Vipeholm studies (Gustafsson *et al.*, 1954) could demonstrate that such high sucrose concentrations can indeed occur in the plaque after consumption of sweets. The *in vitro* cultivation of *S. mutans* can benefit from the toleration of these high concentrations of sucrose by using this effect as a selective factor (Gehring, 1972; Gold, Jordan and van Houte, 1973; Ikeda and Sandham, 1972).

If high amounts of carbohydrates are put at the disposal of the plaque, the high percentage of microorganisms it contains will producc relatively large amounts of acids within minutes. The acids thus formed in the plaque take a comparatively long time to diffuse into the neutralising saliva. Such microbial metabolic end products can accumulate within the plaque with the result that the more acids

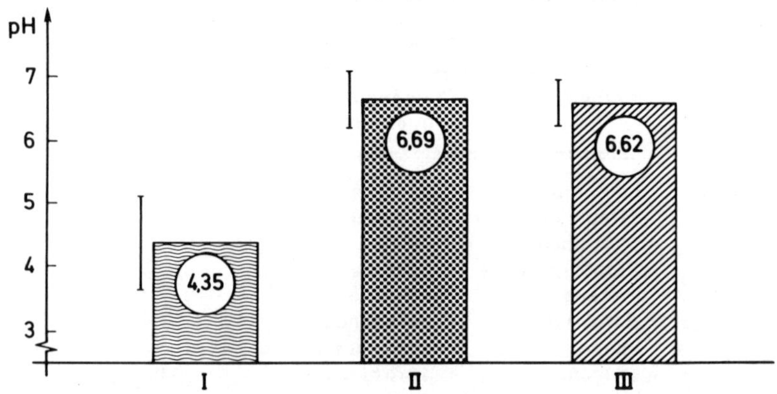

FIG. 3. Acid production by streptococci of the oral cavity in a culture medium with 40% sucrose. I, *S. mutans* (mean values and S.D. of 100 strains; II, *S. sanguis* (40 strains); III, *S. salivarius* (25 strains).

are formed, the longer and stronger will be the attack on the dental enamel. In this case the danger is great that carious lesions will occur before remineralisation can stop this development. The calcium phosphate in the enamel is only slightly soluble in pH neutral or slightly acidic conditions, but the solubility increases with lower pH values with a critical limit for demineralisation of the enamel at pH = 5·5–5·0.

However, the following events can reduce the harmful potential of the acids: diffusion of acids into the saliva; the neutralising effect of saliva and plaque buffers; the neutralising effect of the amines and ammonia formed in the plaque; and the formation of weak acids such as acetic or propionic acid instead of strong acids such as lactic acid.

Synthesis of Extracellular Polysaccharides

A further factor in the aetiology of caries, next to microbial acid production, is the ability to synthesise extracellular polysaccharides from sucrose, a characteristic not only of *S. mutans* but also of *S. sanguis*, *S. salivarius* and *Lactobacillus casei*. *S. sanguis* can be found in relatively great numbers in nearly all plaque samples, and seems to be just as typical an inhabitant of plaque as *S. mutans*. *S. salivarius* on the other hand, preferentially occurs on the tongue and forms the main group of streptococci in the saliva of humans.

While *S. mutans* and *S. sanguis* synthesise polyglucans from sucrose, *S. salivarius* produces mainly polyfructans. The synthesis of these polysaccharides requires free energy, because they represent energy-rich compounds. The free energy is supplied by the unusual energy-rich glycosidic bond of the disaccharide sucrose. The extracellular substances produced, above all, by *S. mutans* have particular physico-chemical properties which are responsible for several functions in caries aetiology. In addition to the cohesive interactions of the bacterial component, such polyglucans constitute an essential part of the plaque material. The sticky nature of the water-insoluble glucans presents *S. mutans* with a selective advantage during colonisation of the teeth surfaces. This advantage can lead to initial caries. The likewise synthesised water-soluble glucans, especially polyfructans, can be metabolised to acids by the microorganisms. They serve as a carbohydrate reserve that can be utilised for the formation of acids in times of little or no food uptake by the host, e.g. at night. Furthermore, the water-insoluble glucans also function as a diffusion barrier and the organic acids produced are kept back in the plaque, leading to a further prolongation of the effect of the acid on the dental enamel. The glucans formed by *S. sanguis*, as well as the fructans, are not sticky and, together with the fact that *S. sanguis* produces less acids, e.g. from sucrose than *S. mutans*, this could explain their low cariogenicity, although they can almost always be found in considerable amounts in plaque.

From all that has been presented so far, it can be concluded that the aetiology of caries depends on a number of factors. A simple diagram (Fig. 4) summarises these factors. The nutritional habits of the host, especially the amount and frequency of sugar con-

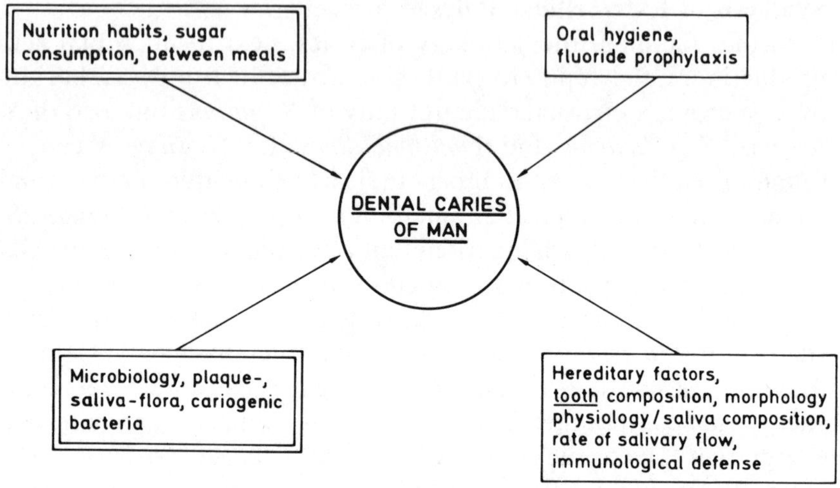

FIG. 4. Some aetiological factors of dental caries: multifactorial process.

sumption, are just as important as the microbiological situation. The more food uptake occurs, the more nutritive substances will be available to the microorganisms in the oral cavity. On the other hand, a well practised oral hygiene can stop the caries process, since it eliminates the nutritive substances on which the oral microflora otherwise thrive. The origin and progress of this illness furthermore depend on genetic factors, on the immunological situation and further physiological factors of the host.

Caries Prevention by Sugar Substitutes

Various defensive measures against the many causes of the caries process can be applied. One of these many possibilities of controlling caries is given by making use of sugar substitutes. They have been studied to a growing extent in recent years due to the following considerations: if in sweets, cariogenic sucrose is substituted by other suitable sweet substances having low or no cariogenic properties at all, then the sweets in question should prove less cariogenic or non-cariogenic. A particular interest has been taken in one of these sugar substitutes, due also to its other favourable qualities: xylitol.

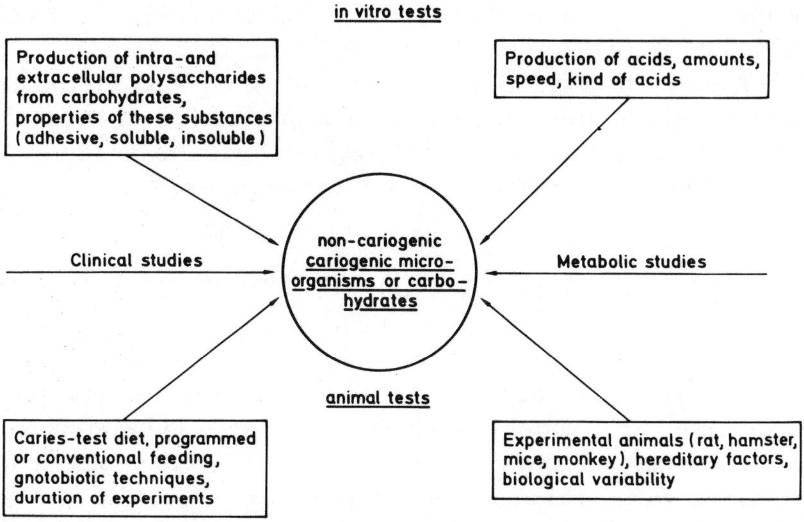

FIG. 5. Some factors responsible for the cariogenicity of microorganisms or carbohydrates—cariogenicity tests.

A number of factors have to be considered in order to estimate to what extent such a substance as xylitol has cariogenic properties. The most important factor concerns the microbial degradation of the substrate to acids—the decisive parameters being the nature, quantity and rate of acid production. Another factor is the microbial synthesis of extra- or intra-cellular polysaccharides from such substances and eventually the consideration of their properties. Like extracellular polysaccharides, the intracellular polysaccharides can also function as a carbohydrate reserve if other nutrients are missing. They can be degraded to organic acids, and this can lead to a continuous effect of the acids, above all during the night. Problems of this sort can be examined by *in vitro* tests, and by testing the cariogenicity in animal experiments. This applies to sugar substitutes as well as to cariogenic properties of microorganisms. Some of the factors in question in cariogenicity tests are presented in Fig. 5. They include clinical and physiological investigations of sugar substitutes, which eventually decide whether these substitutes can be used in caries prophylaxis of humans. Several symposia have dealt in detail with special topics concerned with this group of problems (Ritzel and Brubacher, 1976; Zöllner and Heuckenkamp, 1975; Zucker-Symposion, Würzburg, 1971 and 1977).

Numerous investigations have been carried out with xylitol, and virtually all questions in connection with the application of this sugar substitute in caries prophylaxis have been attended to in the 'Turku Sugar Studies' (Scheinin and Mäkinen, 1975).

PLAQUE MICROFLORA AND XYLITOL (TURKU SUGAR STUDIES)

Over 100 patients participated in the Turku studies. This group was divided into three sections, of which one received food with sucrose, the second with fructose and the third with xylitol. Part of the microbiological investigations concentrated especially on the occurrence of *S. mutans* in the plaque of test persons. At the beginning of the studies it was known that *S. mutans* was not able to metabolise xylitol, and it could be expected that its occurrence would be reduced in the xylitol group. But this would not be the case if a novel ability to metabolise xylitol occurred due to mutation or adaptation of the *S. mutans* organism under the continuous influence of the xylitol diet. Another possibility would be the selection of a few *S. mutans* organisms, previously present in the plaque, with the capability to metabolise xylitol. The number of these organisms should grow and the mixed flora of the plaque in the

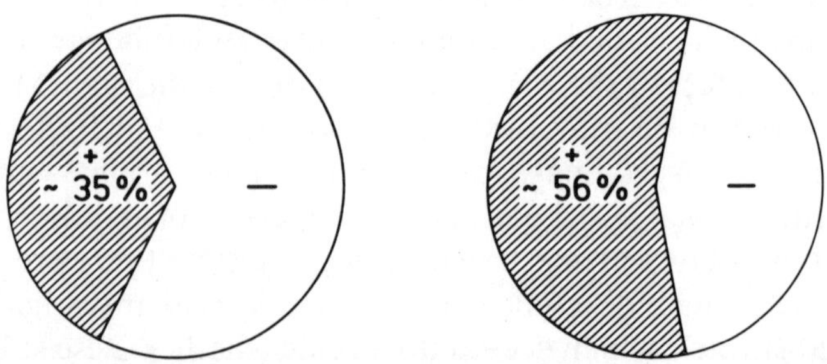

FIG. 6. Distribution of subjects in the xylitol group (left) and the pooled fructose and sucrose groups (right) exhibiting oral growth of *Streptococcus mutans*.

xylitol group should show higher xylitol-degrading activities, after two years, than at the beginning of the experiments.

The results of these experiments indeed demonstrate that the occurrence of *S. mutans* in the xylitol group is significantly lower than in the fructose and sucrose groups (Fig. 6). None of the numerous *S. mutans* strains isolated from plaque samples had acquired xylitol-fermenting abilities. At the end of the experiments not even the mixed flora of the plaque of all three groups could metabolise xylitol to acids in amounts worth mentioning. On the other hand, the acid production from sucrose and fructose reached pH 4–5 after 24 h culture time, and from sorbitol pH 5·5 after 7 days (Fig. 7).

Xylitol-Fermenting Microorganisms

Although the pH values of acids produced from xylitol by the plaque flora were almost never below pH 6·5, there were a few incidents when the pH was 5·5. They occurred sporadically in all three groups, but at the end of the experiments by no means more often in the xylitol group. A few xylitol-fermenting bacterial strains were isolated from these rare cultures of plaque flora cultivated on xylitol. They could be identified as streptococci, but it was not possible to associate them to one of the known serological groups A–S of streptococci (Gehring, 1974). These strains, and similar streptococci that could be isolated from the oral cavity of rats, show the characteristics of the serological streptococci group Q. This is the only serogroup of streptococci in which a positive xylitol degradation has been described: in this case for *Streptococcus avium*, an organism mainly present in faeces of chickens and occasionally in humans, dogs and pigs (Bergey, 1974). Xylitol can be found in plants, and its possible microbial fermentation seems to be a biological necessity, as it must be metabolised in some natural way. Next to streptococci, organisms of the family Enterobacteriaceae from the oral cavity of rats, and not yet identified Gram-negative bacilli from the oral cavity of the hamster show xylitol-fermenting abilities (Gehring, 1974).

Bergey (1974) states that xylitol can also be fermented by a single lactobacillus species, *Lactobacillus salivarius* subsp. *salivarius*, but

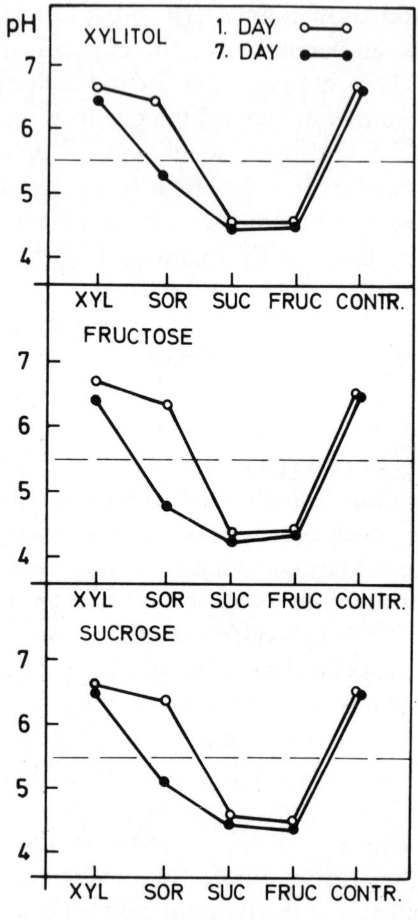

FIG. 7. Mean pH values after incubation for 1 and 7 days in the presence of 1% xylitol, sorbitol, sucrose or fructose. Inoculation with individual plaque samples after 12 months of sugar diets.

this statement could so far not be confirmed. Originally this species was isolated from the mouth and intestinal tracts of the hamster, the mouth of man, and the intestinal tract of the hen. Furthermore, the ability to metabolise xylitol has been described for *Staphylococcus saprophyticus* (Schleifer and Kloos, 1975). According to this meagre information the ability to metabolise xylitol compared with the microbial utilisation of hexoses seems to be rather rare in the world of microorganisms, particularly in the oral cavity of humans. In Fig. 8 this ability of some microorganisms to metabolise xylitol and produce organic acids is compared to the fermentation of other sugar substitutes and sucrose. Additional information on the synthesis of extracellular polysaccharides is also supplied.

Low Acid Production by Fermentation of Xylitol

Acid production from xylitol is low. The diagrams of the pH curves in culture media with different sugar substitutes confirm this fact (Fig. 9). It can be seen that the degradation of the sugar substitutes by xylitol-fermenting streptococci from dental plaque of humans is achieved most rapidly and most intensively with fructose and slowest and weakest with xylitol. Similar results are obtained with the other xylitol-fermenting species mentioned above.

Furthermore, it was of interest to know whether the xylitol-fermenting organisms could acquire the ability of faster and stronger acid production. The degradation of xylitol by a strain of streptococcus originally isolated from the oral cavity of the rat was determined, in the first instance immediately after isolation, and then again after two years. During this time it was exclusively cultivated on a culture medium containing xylitol. The results of this experiment are compared with the corresponding sorbitol, mannitol and fructose degradations in Fig. 10. The graphs of the pH decrease are practically unchanged. Acid production from xylitol thus did not increase after two years of possible adaptation. These results seem to imply that the fermentation of xylitol by xylitol-fermenting organisms is genetically fixed. The degradation of xylitol takes place in metabolic steps that produce less acid and/or other acids than the fermentation of fructose or mannitol and sorbitol. The adaptation of xylitol fermentation in the above-mentioned manner is thus probably not possible and from a biological point of view not necessary.

These circumstances seem to be the decisive reason for the lack of success of the small number of the xylitol-fermenting streptococci in the plaque during the xylitol diet as against the rest of the plaque flora in the Turku sugar studies. In spite of the xylitol diet, sufficient

Microorganisms	Carbohydrates					Synthesis of extracellular polysaccharides
Genus and species	SUCROSE	FRUCTOSE	SORBITOL	L-SORBOSE	XYLITOL	
Streptococcus mutans	+	+	+	−	−	+
" sanguis	+	+	−	−	−	+
" salivarius	+	+	−	−	−	+
" species, not classifiable	(+)	+	+	+	(+)	−
" avium, sero-group Q	+	+	+	+	(+)	−
Lactobacillus casei	(+)	+	+	+	−	+
" salivarius subsp. salivarius	+	+	−	−	±?	
Proteus morganii (rat)	−	+	±	±	(+)	
Staphylococcus saprophyticus	+	±	−	−	±	
Unknown, gram-negative rods (hamster)	+	+	−	−	(+)	

FIG. 8. Microbial acid formation during the breakdown of some carbohydrates and synthesis of extracellular polysaccharides from sucrose. +, Acid formed; ±, acid formation, some strains positive and some negative; (+), slight acid formation.

alternative energy sources were possibly still present. They could be utilised by the mixed flora of the plaque better than xylitol. This also seems to be the case with *S. mutans*. Although no sucrose—the optimum substrate for *S. mutans*—was available in the xylitol group, it could not be suppressed completely in the plaque.

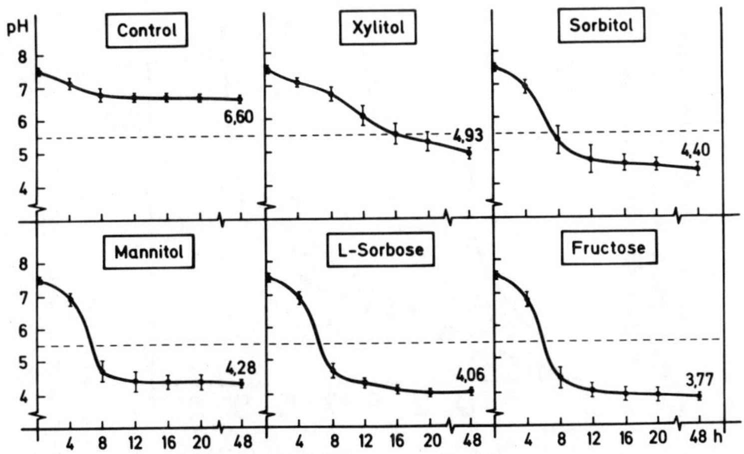

FIG. 9. The effect of various carbohydrates (1 %) on pH of Phenol red broth base (DIFCO) by xylitol-fermenting streptococci species isolated from plaque material of humans. The mean values, end pH values and S.D. of 10 strains. (Turku sugar study).

The significance of *S. mutans* for the aetiology of caries should not be considered isolated from the mass of the other organisms in the plaque. Otherwise, it could happen again, as some decades ago with lactobacilli, that one microorganism will be overrated. At that time, lactobacilli were looked upon as the actual agent of dental caries.

The enormous reduction of the caries rate in the xylitol group of the Turku sugar studies during the two years, is the result of a remarkable reduction of the acid potential in the plaque of test persons. This can be explained by the lack of easily and rapidly fermentable carbohydrates, and the great number of organisms in the plaque flora not able to ferment xylitol. On the other hand, the sporadically appearing xylitol-fermenting organisms, with this weak acid production, had no chance to succeed in the microbial plaque

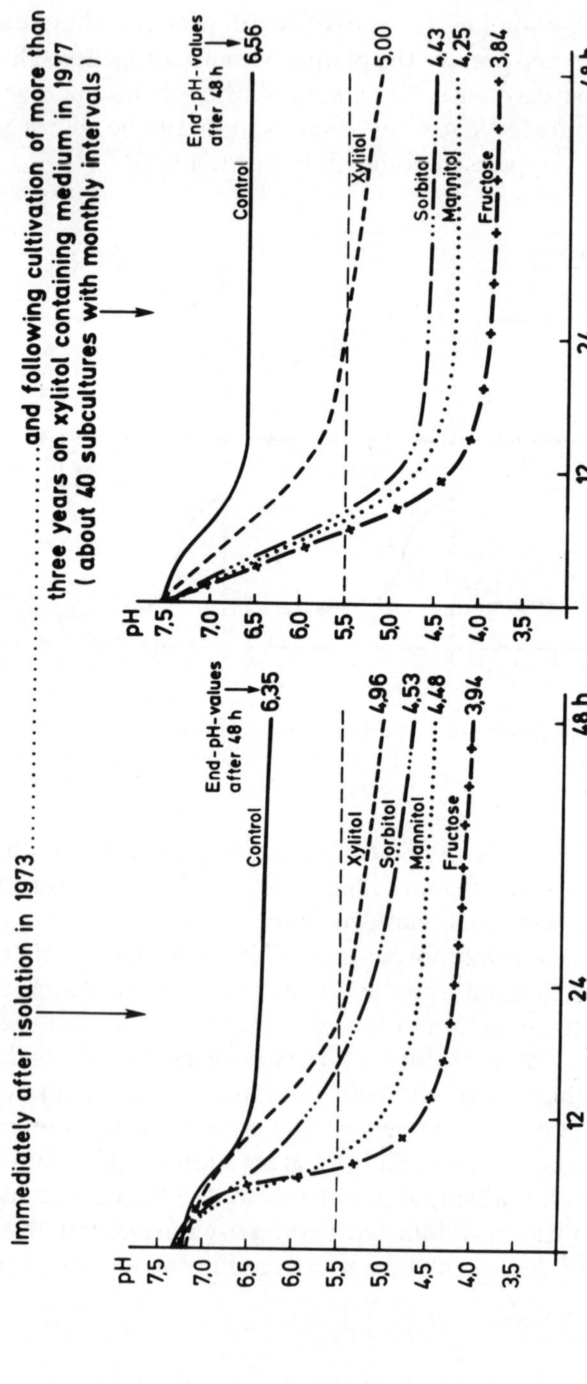

FIG. 10. The effect of various carbohydrates (1 %) on pH of Phenol red broth base (DIFCO) by xylitol-degrading streptococcus species (RX1) isolated from the oral cavity of the rat.

system, with this metabolic performance (Gehring *et al.*, 1975). It is, furthermore, of interest to note that the number of lactobacilli in the xylitol group was reduced considerably (Larmas, Mäkinen and Scheinin, 1975). Usually these organisms occur in the oral cavity of humans in great numbers only in locally limited regions of low pH values, due to the missing rinsing effect of the saliva over long periods of time. This can happen at different retention sites, for instance in caries lesions, but in the xylitol group the number of such caries incidents was reduced to a large extent.

Cariogenicity in Animal Experiments

The determinations of the cariogenicity of sugar substitutes or of certain microorganisms by *in vitro* tests, were complemented by animal experiments. They are absolutely necessary to solve some of the problems. The animal preferably used in these experiments is the rat, but hamsters and monkeys are also suitable. The factors responsible for the origin of caries and the histo-pathological progress in rats and hamsters are very similar to those in humans. Although it can thus be permitted to compare and transfer the results of animal experiments to the situation in humans, these results should nevertheless be interpreted with great care. Caries develop on the molars of experimental animals within a few weeks, irrespective of conventional or programmed uptake of cariogenic food, mainly composed of sucrose. The possibilities of conducting the experiments under controlled conditions and over relatively short periods are great advantages of animal experiments. The evaluation of caries incidence is carried out with the help of histological methods on stained serial sections of the teeth according to an internationally accepted evaluation scheme (König, 1966). Such animal experiments could prove that xylitol is non-cariogenic (Gehring and Karle, 1975; Grunberg, Beskid and Brin, 1973; Karle and Büttner, 1971; Karle and Gehring, 1975; Mühlemann, Regolati and Marthaler, 1970). The results of one of these experiments is shown in Fig. 11. The caries rate after six weeks of programmed feeding of sucrose, sorbitol and xylitol chocolates was determined. The data of the experiment demonstrate quite clearly that xylitol is a non-cariogenic substance. It does not increase the caries rate of the

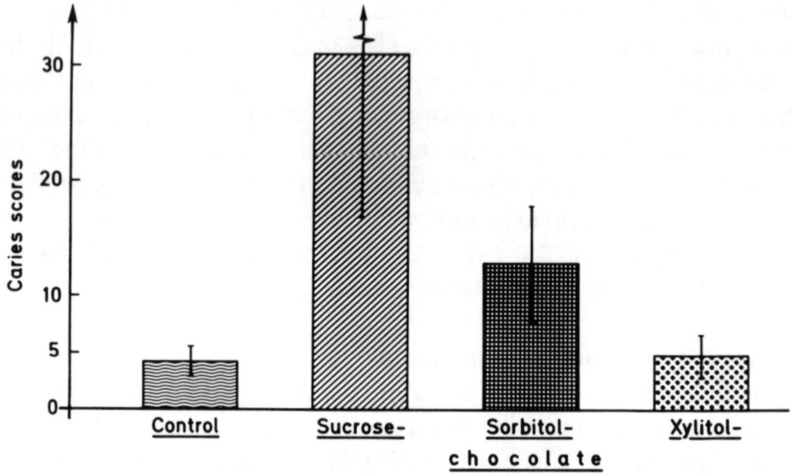

FIG. 11. Caries incidence on the lower jaws of rats after programmed feeding of various chocolates. Mean values and S.D.

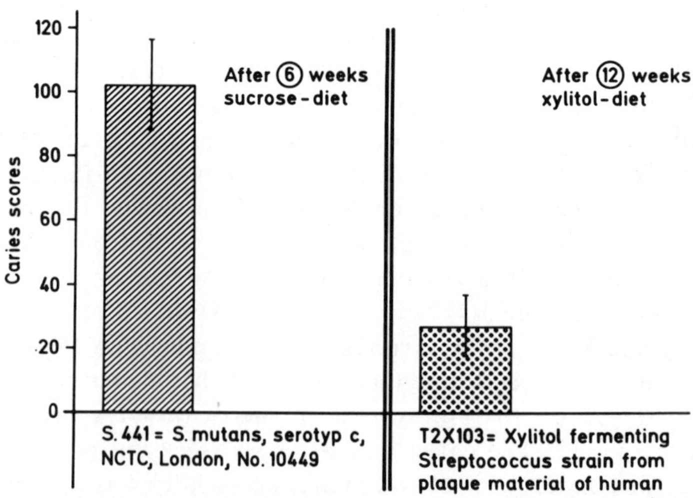

FIG. 12. Comparison of caries scores produced by two streptococci strains in germ-free rats after sucrose or xylitol diet feeding. Mean values and S.D.

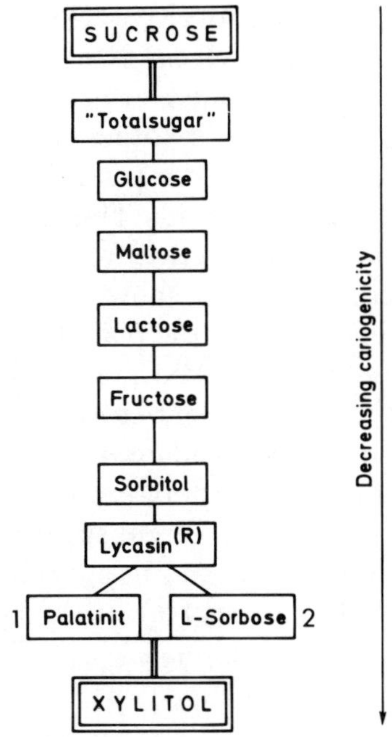

FIG. 13. Cariogenicity of some sugars and sugar substitutes. Palatinit® is glucopyranosido-1,6-sorbitol:glucopyranosido-1,6-mannitol (1:1). 1, Karle and Gehring, 1978; 2, Gehring and Karle, 1977.

animals in the control group. Sorbitol was also significantly less cariogenic compared with sucrose, the most cariogenic substance of this experiment. But sorbitol also had a significantly higher cariogenicity than xylitol.

Since a few xylitol-fermenting strains of streptococci were isolated, the question arose of whether these organisms would show cariogenic properties, if xylitol was supplied. Appropriate experiments were conducted with germ-free rats (Karle and Gehring, 1976) and Fig. 12 presents the results of these experiments with germ-free rats after monoassociation with a xylitol-fermenting strain of streptococci isolated from human plaque material. They were fed on a xylitol diet. For reasons of comparison, a strain of *S. mutans* after feeding with sucrose is added. After six weeks this *S. mutans* strain had caries values near the maximum that can be reached. In contrast, even after twelve weeks, the xylitol-fermenting strains had low caries rates, which can practically be neglected. This result is most remarkable, because in germ-free experiments, the strains can develop, uninhibited by other rival organisms and with sole utilisation of the available substrate. Nevertheless, no caries worth mentioning could be found. This can without doubt be attributed to the weak acid production from fermentation of xylitol. In addition, this experiment clearly shows that even the prevalence of such xylitol-fermenting germs in the plaque of the xylitol group of the Turku sugar studies never evoked a real danger of caries. For the total evaluation of the known sugar substitutes according to their caries-prophylactic possibilities and taking into consideration all the facts presented, xylitol without doubt is the best sugar substitute. A review of the evaluation of the different sugar substitutes according to the present knowledge is given in Fig. 13. It takes into account the published results of the literature of *in vitro* and *in vivo* investigations of humans and experimental animals.

REFERENCES

Bergey, D. H. (1974) *Bergey's Manual of Determinative Bacteriology*. Baltimore: Williams & Wilkins.

Burnett, G. W. & Scherp, H. W. (1968) *Oral Microbiology and Infectious Disease*. Baltimore: Williams & Wilkins.

Carlsson, J. (1967) Presence of various types of non-haemolytic streptococci in dental plaque and in other sites of the oral cavity in man. *Odontol. Revy.*, **18,** 55–74.

Clarke, J. K. (1924) On the bacterial factor in the aetiology of dental caries. *Brit. J. Exp. Path.*, **5,** 141–7.

Ellen, R. P. (1976) Microbiological assays for dental caries and periodontal disease susceptibility. *Oral Science Rev.*, **8,** 3–23.

Gehring, F. (1972) *Extrazelluläre Polysaccharide bildende Streptokokken aus Zahnplaques und ihre Beziehung zur Zahnkaries*. München: Carl Hanser.

Gehring, F. (1974). Mikrobiologische Aspekte zur Kariogenität von Zuckern und Zuckeraustauschstoffen. *Dtsch. zahnärztl. Z.*, **29,** 769–71.

Gehring, F. (1977) Mikrobiologische Untersuchungen im Rahmen der 'Turku sugar studies' *Dtsch. zahnärztl. Z.*, **32,** Suppl. 1, 84–8.

Gehring, F. & Karle, E. J. (1975) Der Zuckeraustauschstoff Xylit und seine Bedeutung in der Kariesprophylaxe. *Süsswaren*, **19,** 330–40.

Gehring, F. & Karle, E. J. (1977) Cariogenicity of 'L-sorbose compared with xylitol and sucrose' in animal studies. *ORCA Congress Abstracts*, Megeve, 1977, Abstract No. 71.

Gehring, F., Mäkinen, K. K., Larmas, M. & Scheinin, A. (1975) Turku sugar studies. X. Occurrence of polysaccharide-forming streptococci and ability of the mixed plaque microbiota to ferment various carbohydrates. *Acta Odont. Scand.*, Suppl. **70,** 223–37.

Gibbons, R. J. & van Houte, J. (1975) Dental caries. *Ann. Rev. Med.*, **26,** 121–36.

Gold, O. G., Jordan, H. V. & van Houte, J. (1973) A selective medium for *Streptococcus mutans*. *Archs. Oral Biol.*, **18,** 1357–64.

Grunberg, E., Beskid, G. & Brin, M. (1973) Xylitol and dental caries. *Int. J. Vit. Nutr. Res.*, **43,** 227–32.

Gustafsson, B. E., Quensel, C. E., Swenander Lanke, L., Lundquist, C., Grahnen, H., Bonow, B. E. & Krasse, B. (1954) The effect of different levels of carbohydrate intake on caries activity in 436 individuals observed for five years. *Acta Odont. Scand.*, **11**.

Ikeda, T. & Sandham, H. J. (1972) A high sucrose medium for the identification of *Streptococcus mutans*. *Archs. Oral Biol.*, **17,** 781–3.

Karle, E. & Büttner, W. (1971) Kariesbefall im Tierversuch nach Verabreichung von Sorbit, Xylit, Lycasin und Calciumsaccharosephosphat. *Dtsch. zahnärztl. Z.*, **26,** 1097–105.

Karle, E. & Gehring, F. (1975) Wirkung der Zuckeraustauschstoffe Fruktose, Sorbit und Xylit auf Kariesbefall und Plaqueflora der Ratte. *Dtsch. zahnärztl. Z.*, **30,** 356–63.

Karle, E. J. & Gehring, F. (1976) Zur Kariogenität xylitabbauender Keime im gnotobiotischen Versuch. *Dtsch. zahnärztl. Z.*, **31,** 22–5.

Karle, E. J. & Gehring, F. (1978) Palatinit®—ein neuer Zuckeraustauschstoff und seine kariesprophylaktische Beurteilung. *Dtsch. zahnärztl. Z.*, **33,** (in press).

König, K. G. (1966) *Möglichkeiten der Kariesprophylaxe beim Menschen und ihre Untersuchung im kurzfristigen Rattenexperiment*. Bern & Stuttgart: Hans Huber.

Larmas, M., Mäkinen, K. K. & Scheinin, A. (1975) Turku sugar studies. VIII. Principal microbiological findings. *Acta Odont. Scand.*, **33,** Suppl. 70, 173–216.

McHugh, W. D. (1970) *Dental Plaque*. University Court of the University of Dundee.

Miller, W. D. (1890) *The Micro-Organisms of the Human Mouth*. Philadelphia: S. S. White Dental Manufacturing Company.

Mühlemann, H. R., Regolati, B. & Marthaler, T. M. (1970) The effect on rat fissure caries of xylitol and sorbitol. *Helv. Odont. Acta*, **14,** 48–50.

Nolte, W. A. (1968 & 1973) *Oral Microbiology*. St. Louis: C. V. Mosby.

Ritzel, G. & Brubacher, G. (1976) *Monosaccharides and Polyalcohols in Nutrition, Therapy and Dietetics*. Definierte Monosaccharide und Zuckeralkohole in Ernährung, Therapie und Diätetik. Beiheft zur Internationalen Zeitschrift für Vitamin- und Ernährungsforschung, Nr. 15. Bern–Stuttgart–Wien: Hans Huber.

Scheinin, A. & Mäkinen, K. K. (1975) Turku sugar studies. I–XX. *Acta Odont. Scand.*, **33,** Suppl. 70.

Schleifer, K. H. & Kloos, W. E. (1975) Isolation and characterization of Staphylococci from human skin. I. Amended descriptions of *Staphylococcus saprophyticus* and descriptions of three new species: *Staphylococcus cohnii*, *Staphylococcus haemolyticus*, and *Staphylococcus xylosus*. *Int. J. Syst. Bacteriol.*, **25,** 50–61.

Skinner, F. A. & Carr, J. G. (1974) *The Normal Microbial Flora of Man*. London & New York: Academic Press.

Stiles, H. M., Lösche, W. J. & O'Brien, T. C. (1976) *Microbial Aspects of Dental Caries*. Vols. I, II & III. Washington D.C. & London: Information Retrieval.

Zöllner, N. & Heuckenkamp, P.-U. (1975) Zucker und Zuckeraustauschstoffe [Sugars and sugar substitutes]. *Nutr. Metabol.*, **18,** Suppl. 1, 1–266.

Zucker-Symposion I, Würzburg (1971) *Dtsch. zahnärztl. Z.*, **26,** H.11.

Zucker-Symposion II, Würzburg (1977) *Dtsch. zahnärztl. Z.*, **32,** Suppl. 1.

CHAPTER 8

Behaviour of Xylitol in Isolated Plaque

G. FROSTELL

(*Faculty of Odontology, Lund University, Malmö, Sweden*)

INTRODUCTION

A discussion of the behaviour of xylitol in isolated dental plaque may be divided into some major topics such as fermentation of xylitol, effects of xylitol on bacterial growth, and production of insoluble polysaccharides. However, the discussion may be further subdivided into the following: absence or reduction of fermentation (compared with sucrose); absence or reduced production of insoluble plaque polysaccharides (compared with sucrose); inhibition by xylitol of acid production from glucose or sucrose; inhibition by xylitol of production of insoluble plaque polysaccharides from sucrose; inhibition by xylitol of production of acid from sorbitol or Lycasin ®; bacteriostatic or bactericidal effects of xylitol. There are also other possibilities. Apart from all these different factors, there is also always the question of whether an *adaptation* may take place which would change the behaviour of xylitol.

BEHAVIOUR OF XYLITOL IN DENTAL PLAQUE

Fermentation

It is well known from a number of studies that xylitol is not fermented by the majority of oral microorganisms. However, it is also known that some strains of lactobacilli and streptococci may produce acid from xylitol. Oral rinses with xylitol solutions do not provoke pH decreases in dental plaques as shown by Mühlemann and coworkers and as found in our own studies. These results indicate that if there is any acid production in isolated dental plaque, it should also be very low.

In order to test this hypothesis our group (Ahldén, Birkhed, Edwardsson and Frostell, 1977) has studied acid production in suspensions of dental plaque material in a large number of subjects. The subjects had to refrain from tooth-brushing for two days after which the dental plaque material was removed from the teeth using an instrument. The material was weighed and suspended in a weak phosphate solution. This plaque suspension was then transferred to a constant-temperature glass vessel and oxygen-free nitrogen gas was passed through the suspension. The pH was kept constant automatically by a Radiometer titration machine and the amount of sodium hydroxide required to keep the pH constant was a measure of the acid-production activity. Acid-production activity was expressed as nano-equivalents of acid produced per minute per milligram dental plaque material (wet weight). This method is rather sensitive and exact. In none of twenty subjects could any acid production from xylitol be demonstrated. In a similar series of experiments, acid production from glucose, sorbitol, Lycasin® (Swedish candy quality, Lyckeby Starch Refining Company) and maltitol was determined. A lively acid production was demonstrated from glucose, whereas the acid production from sorbitol and maltitol was about 20–40 % of that from glucose and acid production from Lycasin® was in the order of 50–60 % of that from glucose.

The xylitol experiment was repeated with another group of twenty subjects with similar results. Thus, no acid production could be demonstrated from any of the plaque suspensions on contact with xylitol.

In order to test whether an *adaptation* to xylitol could take place as a result of frequent consumption of pastilles containing xylitol, twenty subjects were given two 1-g pastilles containing 50 % xylitol four times a day for three months. After this experimental period the acid-production experiments were repeated. No acid production from xylitol could be demonstrated in any of the suspensions. The subjects also rinsed their mouths with a 50% solution of xylitol before and after the three-month experimental period, and the pH changes were determined according to a method published by Frostell (1970). There was no statistical difference in the pH changes before and after the experimental period. There was also no significant difference in the plaque weights before and after this experimental period nor was there any acid production from xylitol before or after this adaptation period. In all, we have now made about a hundred titration experiments with dental-plaque material from about sixty subjects, and in no case have we been able to demonstrate any acid production at all from xylitol. Thus, acid production from xylitol in isolated plaque, if any, is very low. Adaptation to acid production could not be demonstrated after three months of repeated daily consumption of xylitol. However, in our opinion we cannot as yet exclude the possibility of a certain adaptation to utilisation of xylitol by dental-plaque material after frequent consumption of this substance for a long time either by selection, induction or even mutation.

Production of Insoluble Plaque Polysaccharides

It is well known that polysaccharides may be produced in isolated plaque from sucrose, glucose and fructose. Production of insoluble plaque polysaccharides from sucrose is believed to be intimately correlated with plaque formation and the caries process. This mechanism is known to be of special importance for smooth-surface and gingival caries. Animal experiments as well as experiments with humans indicate a reduced plaque formation on frequent consumption of xylitol in comparison with frequent consumption of sucrose. There are, to my knowledge, no studies on polysaccharide production from xylitol by isolated plaque. It has been shown by Mäkinen (1972) that *Streptococcus mutans* (strain Ingbritt) does not utilise xylitol at all, and it seems rather unlikely that insoluble polysaccharides are produced to any significant extent from xylitol.

However, this problem could easily be studied in isolated plaque also.

Possible Inhibition of Acid Production from Glucose or Sucrose

There is no indication that xylitol inhibits acid production from glucose or sucrose as revealed by rinsing experiments and pH determinations in dental plaque *in vivo*. In our own titration experiments with plaque suspensions we have been able to show that even rather high concentrations of xylitol do not influence acid production from glucose. However, we have not yet carried out a more extensive production inhibition study.

Possible Inhibition of Production of Insoluble Plaque Polysaccharides

Knuutila and Mäkinen (1975) recently reported that low concentrations of xylitol activated polysaccharide synthesising enzymes whereas higher concentrations (100 mg/ml) partly inhibited them. There is, of course, a possibility that in the presence of xylitol, the production of polysaccharides from sucrose could be inhibited by xylitol in suspensions of isolated plaque.

Possible Inhibition of Production of Acids from Sorbitol or Lycasin®

Since sorbitol and Lycasin® and many other sugar substitutes are much less expensive than xylitol, it would be of great interest to see whether acid production, if any, from these substitutes, would be influenced by the presence of xylitol. Although there have been suggestions in this direction there is still no definite proof of such an action of xylitol in isolated dental plaque.

Possible Bacteriostatic or Bactericidal Effects

Since it was found by Mäkinen in 1972 and Knuutila and Mäkinen (1975) that growth of microorganisms in Trypticase Soy Broth was to some extent inhibited in the presence of xylitol, it may be suspected that xylitol may have bacteriostatic or bactericidal effects in suspensions of isolated plaque. However, to my knowledge there are no results indicating a bacteriostatic or bactericidal effect of xylitol in suspensions of isolated plaque. This hypothesis could easily be tested. Mäkinen's results indicate that the inhibiting effect of xylitol is reduced if the cells are stored in the presence of xylitol for some months, which indicates that the organisms to some extent may either adapt themselves to the action of xylitol or utilise some other substrate molecule, for example a protein.

Our own bacteriological studies on isolated dental plaque have revealed occasional strains of lactobacilli and streptococci, which possess the capacity to ferment xylitol. A study of the composition of the plaque flora before and after a three-month period of frequent consumption of xylitol could not demonstrate any significant changes in the flora as a whole or in the occurrence of such occasional xylitol-fermenting strains.

The knowledge of our group in this field may be summarised as follows: acid production from xylitol in isolated plaque, if any, is very low. There was no adaptation to fermentation of xylitol after three months' frequent consumption of the substance. There is no indication that production of insoluble plaque polysaccharides from sucrose may perhaps be activated by low concentrations of xylitol and partly inhibited by higher concentrations. There is no definite proof of an inhibition of acid production from sorbitol or Lycasin® by xylitol. We are not aware of any pronounced bacteriostatic or bactericidal effects of xylitol in isolated plaque. It may be stated that there is still very much to study concerning the effects of xylitol in isolated plaque.

REFERENCES

Ahldén, M.-L., Birkhed, D., Edwardsson, S. & Frostell, G. (1977) Quantitative determination of the acid production from maltitol, sorbitol and xylitol by streptococci and lactobacilli. In press.

Bramstedt, F. & Trautner, K. (1971) Zuckeraustauschstoffe und Biochemie der Zahnplaques. *Dtsch. Zahnärtzl. Z.*, **11,** 1135–44.

Edwardsson, S. & Birkhed, D. (1977) Effects of 3 months frequent consumption of Lycasin®, maltitol, sorbitol and xylitol on human dental plaque. II. Bacteriological investigations. In press.

Edwardsson, S., Birkhed, D. & Mejàre, B. (1977) Acid production from Lycasin®, maltitol, sorbitol and xylitol by oral streptococci and lactobacilli. In press.

Frostell, G. (1959) A titration method for the determination of acid and alkali production by microorganisms. *Acta Odont. Scand.*, **17,** 441.

Frostell, G. (1964) Quantitative determination of the acid production from different carbohydrates in suspensions of dental plaque material. *Acta Odont. Scand.*, **22,** 457.

Frostell, G. (1970) A method for evaluation of acid potentialities of foods. *Acta Odont. Scand.*, **28,** 599.

Gehring, F. (1971) Saccharose und Zuckeraustauschstoffe im mikrobiologischen Tests. *Dtsch. Zahnärtzl. Z.*, **26,** 116.

Gehring, F., Mäkinen, K. K., Larmas, M. & Scheinin, A. (1975) Occurrence of polysaccharide forming streptococci and ability of the mixed plaque microbiota to ferment various carbohydrates. Turku Sugar Studies. X. *Acta Odont. Scand.*, **33,** Suppl. 70, 223.

Karle, E. & Gehring, F. (1975) Wirkung der Zuckeraustauschstoffe Fruktose, Sorbit und Xylit auf Kariesbefall und Plaqueflora der Ratte. *Dtsch. Zahnärtzl. Z.*, **30,** 356.

Knuutila, M. L. & Mäkinen, K. K. (1975) Effect of xylitol on the growth and metabolism of *Streptococcus mutans. Caries Res.*, **9,** 177.

Mäkinen, K. K. (1972) Enzyme dynamics of a cariogenic streptococcus; the effect of xylitol and sorbitol. *J. Dent. Res.*, **51,** 403.

Mühlemann, H. R. (1969) Zuckerfreie, zahnschonende und nicht-kariogene Bonbons und Süssigkeiten. *Schweiz. Mschr. Zahnheilk.*, **79,** 117.

Mühlemann, H. R. & de Boever, J. (1969) Radiotelemetry of the pH of inter-dental areas exposed to various carbohydrates. In: *Dental Plaque*, ed. W. D. McHugh, pp. 179–86. Edinburgh & London: Livingstone.

CHAPTER 9

Incidence of Clinical and Experimental Caries in Relation to Microbiological and Biochemical Xylitol-Induced Effects

A. SCHEININ
(*Institute of Dentistry, University of Turku, Turku, Finland*)

INTRODUCTION

A massive reduction of the caries increment rate has been observed with total and partial substitution of dietary sucrose by xylitol (Scheinin, Mäkinen and Ylitalo, 1974 and 1975; Scheinin *et al.*, 1975). The very low incidence of dental caries as registered in these trials in relation to xylitol consumption has rapidly stimulated further research in this field. Independent clinical trials and field studies involving a partial substitution of sucrose with xylitol have thus been planned and are shortly to be initiated. So far, however, the bulk of new information has been limited to studies showing the behaviour of xylitol with regard to the following: decomposition influence on the physical properties and chemical composition of plaque and oral fluid, fitness for microbial metabolism, effects in animal studies and metabolic reactions in general.

In view of the association between these observations and the findings on dental caries, the purpose of this contribution is to summarise the essential results from the Turku studies and to interrelate this information with recent microbiological, biochemical and other experimental results.

THE TURKU TWO-YEAR FEEDING STUDY

The trial involved almost complete substitution of sucrose by fructose or xylitol during a period of two years. In addition to the clinical and radiographic registrations of dental caries, biochemical and microbiological investigations were carried out in plaque and saliva and a series of comprehensive blood and urine tests were conducted in order to monitor the general health and to identify side effects occurring in the subjects.

Initially, there were 125 volunteers having a mean age of 27·6 years. The subjects were divided, partly on an individual preference basis, into three groups: the sucrose group comprised thirty five subjects, the fructose group thirty eight and the xylitol group fifty two subjects, the last group being made oversize due to fear of loss of volunteers. During the study ten subjects discontinued or were otherwise excluded. Only in one case was this due to osmotic diarrhoea in relation to xylitol consumption; the other cases were due to difficulties in adhering to the strict dietary regimen, other personal reasons, and in three cases, (two in the sucrose and one in the fructose group) due to excessive caries incidence. The final results are thus based on 115 individuals.

Initially, no significant differences were observed between the groups with regard to age, sex, number of primary and secondary carious surfaces with and without defect, number of filled surfaces and extracted teeth, and the DMFS index (Scheinin, Mäkinen & Ylitalo, 1974). During the trial the dental conditions were registered on eight occasions, the frequent inspections being due to the potential harmful effects of the sugars requiring an early diagnosis and treatment, and enabling participation to be terminated. The scheme used in the diagnosis, classification and treatment of the results is given in Fig. 1. The parameters at the clinical (C1, C2, CS 1, CS 2, F) and radiographic (CR 1, CR 2, CSR 1, CSR 2) registrations were utilised as a basis for calculating the reversals in diagnosis between the examinations, and the resulting net changes. The positive reversals involved a change in diagnosis from an intact surface (0) to C1 or C2, or when considering separately the increment in lesion size, from C1 to C2. A negative reversal indicated a

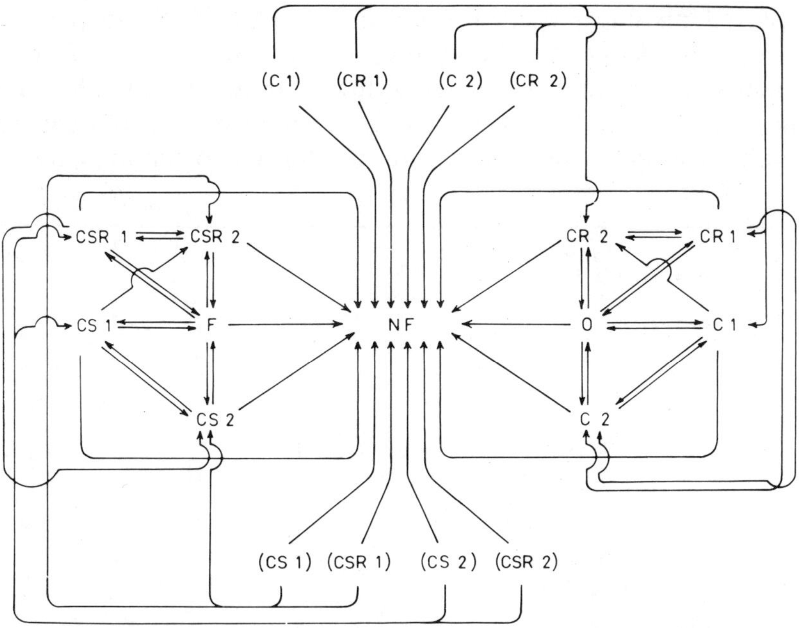

FIG. 1. Development of clinical and radiographic caries reversals. The occurrence of positive and negative reversals is indicated through arrows. The scheme is incomplete regarding development of baseline lesions (in brackets). 0, Intact tooth surface; C, clinical caries reversal; CR, radiographical caries reversal; CS, secondary caries reversal; CSR, radiographical secondary caries reversal. 1, Lesion without defect; 2, lesion with defect. F, Filled surface (at baseline examination); NF, new filled surface; ( ), prestudy carious lesions (at baseline examination).

corresponding change in the opposite direction. Consequently net increment was expressed as new carious surfaces, i.e. positive minus negative reversals (COCSTOC, 1973). The overall comparison of the differences was carried out by using the Kruskal–Wallis test, and the differences between the sugar groups tested for significance with the Mann–Whitney U-test (Bradley, 1968; Chilton, 1967).

According to the clinical and radiographic registrations, a highly significant reduction exceeding 85 % was found in the xylitol group when compared with the sucrose group after one year of the dietary regimen. With fructose, the corresponding reduction was in the order of 25 %. After two years, the increment in the number of decayed, missing and filled tooth surfaces (Δ DMFS index) was 7·2 in the sucrose group, the corresponding values being 3·8 in the fructose group and 0·0 in the xylitol group (Fig. 2). The caries incidence was also expressed in combined quantitative and qualitative terms, by considering in addition to the increment in

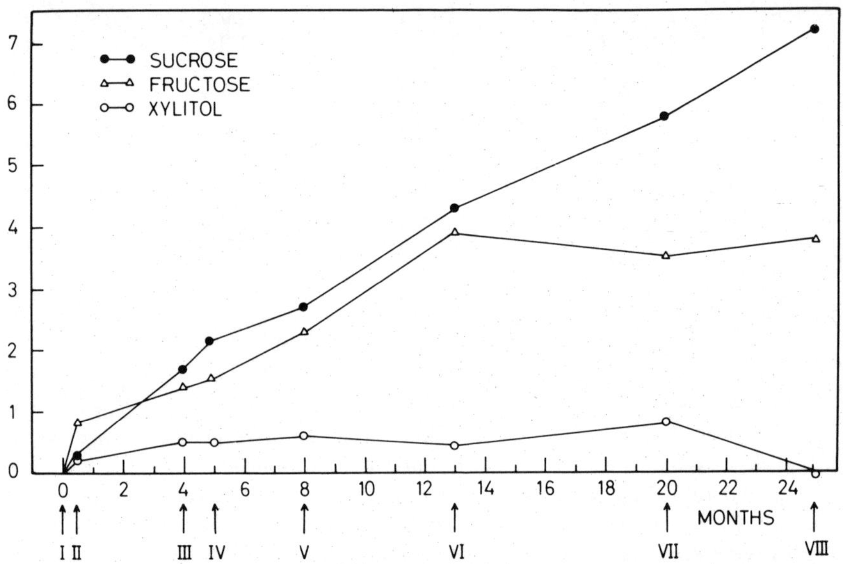

FIG. 2. Increment of decayed, missed and filled tooth surfaces (Δ DMFS index during two years of selective sucrose-, fructose- and xylitol-containing diets). The arrows indicate the clinical and radiographical registrations of dental caries.

terms of the conventional DMFS index, also the changes in lesion size and the secondary caries incidence. The total number of these changes at the various registrations, and the cumulative development is shown in Fig. 3, the differences between the xylitol and sucrose groups being highly significant, but not between the fructose and sucrose groups. Irrespective of the method of expressing the caries increment rate in qualitative, quantitative or combined terms, the reduction in the caries incidence in the xylitol group, as compared with the sucrose group exceeded 85 %, and correspondingly 30 % in the fructose group.

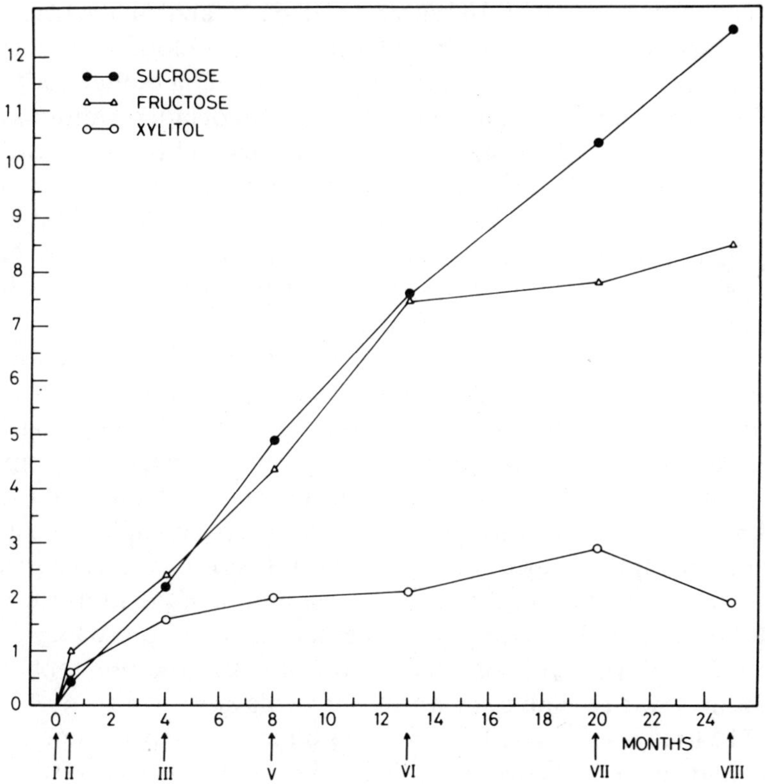

FIG. 3. Cumulative increment of quantitative and qualitative reversals. Sum of increment in number of decayed, missed and filled tooth surfaces (Δ DMFS), all secondary caries reversals (F → CS 1, F → CS 2, F → CSR 1, F → CSR 2), and increase in size of total clinical and radiographic reversals (C 1 ⇄ C 2, CS 1 ⇄ CS 2, CR 1 ⇄ CR 2, CSR 1 ⇄ CSR 2). The arrows indicate the clinical and radiological registrations of dental caries.

THE TURKU ONE-YEAR CHEWING-GUM STUDY

The second study comprised partial substitution, the effects of a sucrose- or xylitol-containing chewing gum being compared during one year. In this study also, the clinical radiographic recordings of caries were supplemented by biochemical and microbiological analyses of plaque and saliva.

The study began with 102 subjects (28 males and 74 females). The mean age was 22·2 years, the subjects being predominantly dental and medical students. Before commencement of the actual study, the volunteers were first examined with regard to dental conditions and dietary habits. The registration of the dental conditions was carried out clinically and radiographically as described earlier (Scheinin, Mäkinen and Ylitalo, 1974). Following initial analysis of the dietary habits during a five-day period in the manner described by Nizel (Nizel, 1972), the subjects were divided fully at random into the future sucrose and xylitol groups.

The subjects were instructed to maintain their regular dietary habits and oral hygiene procedures. An ample supply of sucrose or xylitol chewing gum, each stick weighing 3·0 g and containing 1·5 g of the respective sweetener, was given to each participant. Instructions were further given to consume between three and seven sticks at spaced intervals per day, and to keep a complete record of the consumption during the projected full study year. The dietary habits were further checked before termination of the study during a 14-day period. By taking into account also the preceding 5-day period, information on the frequency of sucrose-containing products was available for a total of 19 days.

The final results were based on 100 subjects, one person not being allowed to enter the assigned xylitol group due to excessive caries incidence, and a further subject being excluded due to lack of cooperation. Initially, there were no significant differences between the groups with regard to the experimental variables, with the exception that the number of missing teeth was slightly, but significantly, higher among the females in the xylitol group than in the sucrose group (Scheinin *et al.*, 1975). It is thought, however, that this feature should have no influence on the results obtained.

The consumption of chewing gum, calculated as number of sticks per day was 4·0 in the sucrose group, and 4·5 in the xylitol group, the difference between the groups not being significant.

The frequency of sucrose consumption, calculated as the sum of the sucrose-containing products ingested in liquid and solid form, at, and between, meals, was 4·24 per subject per day in the sucrose group, and correspondingly 4·94 in the xylitol group. This included sucrose consumed in solid form between meals 1·9 times per day in the sucrose group, the corresponding value being 2·1 in the xylitol group. The values for the above four sub-classes were consistently higher in the xylitol group than in the sucrose group, although the difference was generally not significant.

The cumulative increment of the number of decayed, missing and filled tooth surfaces during the study (Δ DMFS index) at the 6- and

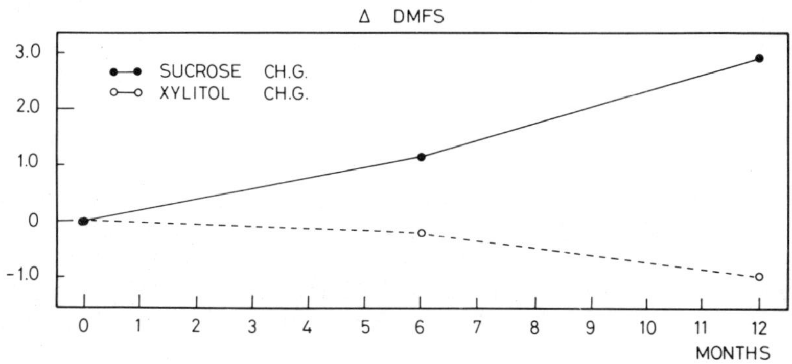

FIG. 4. Cumulative development of decayed, missed and filled tooth surfaces (Δ DMFS index) during 12 months of sucrose or xylitol chewing-gum (CH.G.) intake.

12-month registrations is shown in Fig. 4, the differences between the groups being significant after 6 months ($p = 0.006$) and highly significant after 12 months ($p < 0.001$). This calculation does not provide a complete description of the total quantitative development, particularly as the development of secondary caries remains obscured by those tooth surfaces already being considered as filled. The caries incidence after 12 months, when considering in addition to the increment of decayed, missing and filled tooth surfaces (Δ DMFS), also the secondary caries reversals (CS 1, CS 2, CSR 1 and CSR 2), was 3·76 in the sucrose group, and 0·33 in the xylitol group, this difference being highly significant ($p < 0.001$). The caries incidence may also be expressed in combined quantitative and qualitative terms, by considering in addition to these mean values representing the total quantitative development, also the qualitative development in terms of changes in the lesion size (C 1 ⇄ C 2,

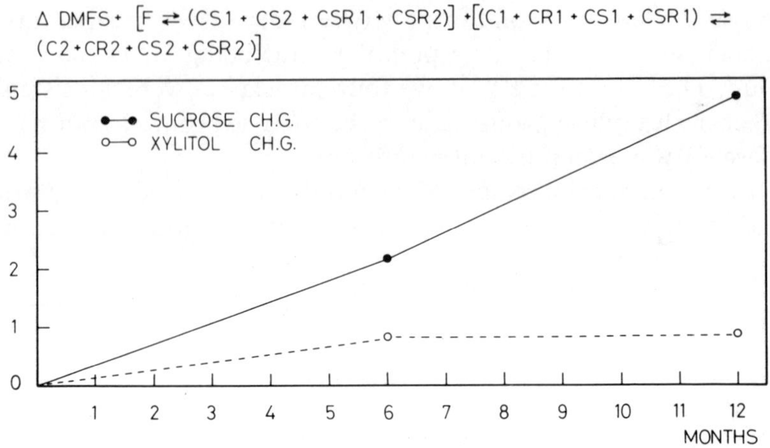

FIG. 5. Cumulative increment of quantitative and qualitative reversals. Sum of increment in number of decayed, missed and filled tooth surfaces (Δ DMFS), all secondary caries reversals (F → CS 1, F → CS 2, F → CSR 1, F → CSR 2), and increase in size of total clinical and radiographic reversals (C 1 ⇄ C 2, CS 1 ⇄ CS 2, CR 1 ⇄ CR 2, CSR 1 ⇄ CSR 2).

CS 1 ⇄ CS 2, CR 1 ⇄ CR 2, CSR 1 ⇄ CSR 2). The development of the latter caries activity index is shown in Fig. 5, the differences between the sugar groups being highly significant after 12 months ($p < 0{\cdot}001$).

FURTHER STUDIES ALLIED TO THE MAIN PROJECTS

Associated studies were initiated parallel to the main trials. These studies involved determinations on periodontal tissues, plaque, saliva, serum and urine in order to observe other accompanying differences on an individual level and between groups. The biochemical, microbiological, periodontal and general methodology was reported separately (Scheinin and Mäkinen, 1975). The corresponding scope of the chewing-gum trial was limited to biochemical and microbiological studies in plaque and saliva.

In the feeding study, the mean individual monthly intake of sucrose, fructose and xylitol was 2·2, 2·1 and 1·5 kg, respectively. Osmotic diarrhoea occurring initially in a number of subjects in the xylitol group gradually disappeared as a phenomenon of adaptation took place (Fig. 6). Later the occurrence of diarrhoea in the xylitol group was of nearly the same frequency as in the sucrose and fructose groups (Mäkinen and Scheinin, 1975A).

The principal biochemical findings on whole saliva and plaque (Mäkinen and Scheinin, 1975B) also included quantification of plaque. Throughout the study, those subjects on the xylitol diet maintained a significantly lower amount of plaque than those on the fructose and sucrose diets. Decreased lactate concentrations of plaque and whole saliva, diminished activity of whole saliva amylase, and reduced hydrolysis rate of sucrose in plaque and whole saliva were observed in relation to xylitol consumption. Increased activity in plaque of certain glycosidases and lactoperoxidase in saliva was associated with the intake of xylitol.

The microbiological observations showed that the dietary regimen did not affect the major microbial categories occurring in plaque or saliva (Larmas *et al.*, 1975). On the other hand, the mean values of viable *Streptococcus mutans* in plaque were lower in the xylitol group than in the other two sugar groups throughout the dietary phase of the study (Gehring *et al.*, 1974 and 1975). Additionally, the geometric and arithmetic means of the colony-forming units on selective Rogosa S.L. and Sabouraud antibiotic agar were significantly lower in the xylitol group than in the fructose and sucrose groups. A reduction of the acidogenic and aciduric oral flora was thus observed particularly in the xylitol group (Larmas, Mäkinen and Scheinin, 1975). During the course of the study, no evidence was obtained of adaptation or mutation enabling acidogenic decomposition of xylitol (Gehring *et al.*, 1975).

A further aim was to determine, using clinical, radiographic and biochemical methods, whether chronic consumption of xylitol could indirectly affect the response reactions of the periodontal tissues. During the diet period, no significant differences among the sugar groups could be established (Paunio *et al.* 1975). It should be noted, however, that periodontal registrations were not carried out at the beginning of the trial, and comparisons with the initial level were thus excluded. In all three sugar groups, the gingival conditions improved toward the end of the study.

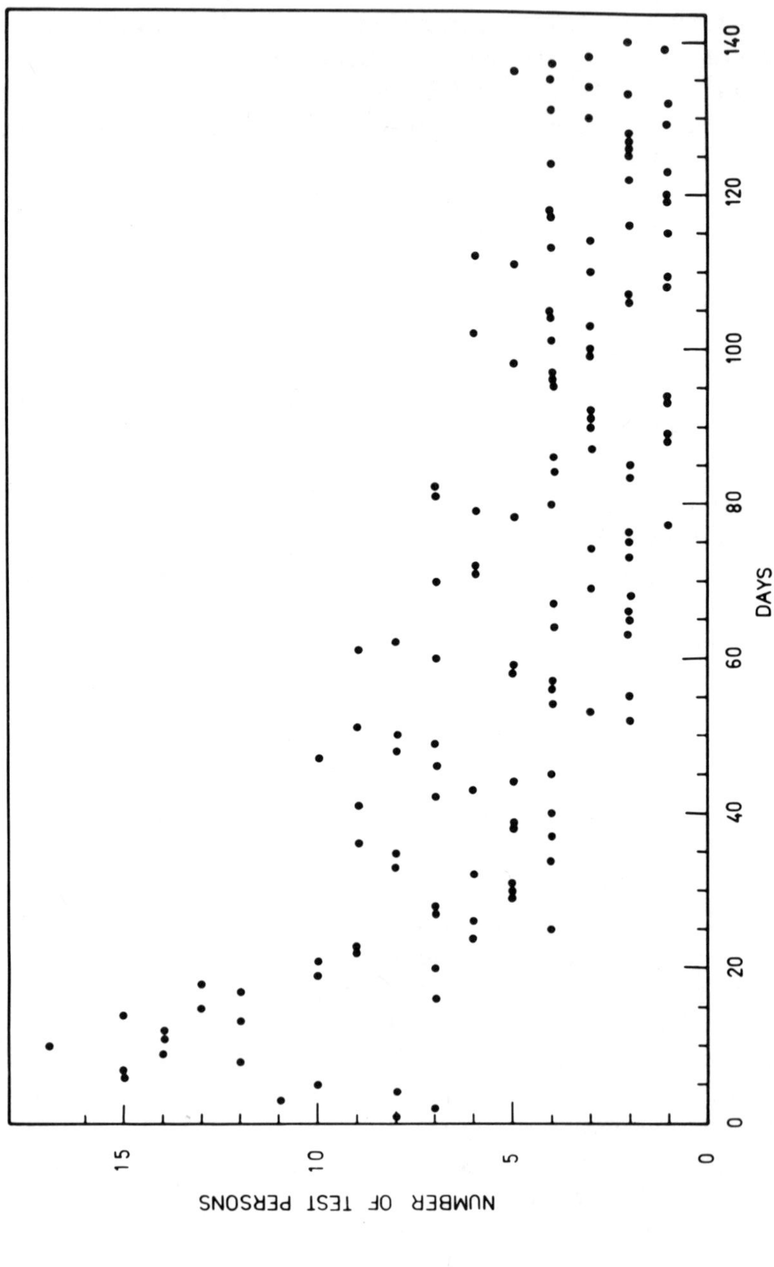

FIG. 6. The occurrence of reported diarrhoea and flatulence-like conditions in the xylitol group during the first 140 days of the two-year feeding study. From Mäkinen and Scheinin (1975A).

During the planning phase of the trial, due consideration was given to the unrealistic nature of the goal of total substitution of sucrose by fructose or xylitol. Recognition of this fact enhanced the importance of the accompanying surveys on the general health of the subjects. The effects of chronic consumption of fructose and xylitol were thus studied with regard to glucose, lipid and urate metabolism (Huttunen *et al.*, 1975), and further metabolic behaviour, including liver-function tests and haematological assays (Mäkinen and Scheinin, 1975C). These studies were expanded by disc electrophoresis of certain serum proteins (Mäkinen, Mielityinen and Scheinin, 1975).

No consistent differences were found in serum triglycerides, glucose, insulin, urate, lactate and pyruvate concentrations, or in the urinary excretion of urate among the experimental groups. Serum cholesterol tended to be lower in the fructose than in the xylitol group, but the difference disappeared when subjects with initial high serum cholesterol in the baseline examination were excluded (Huttunen *et al.*, 1975). Individual and pooled serum samples of the subjects were further analysed for various electrolytes, bilirubin, ascorbate, alkaline and acid phosphatase, amylase, transaminases, lactate dehydrogenase and amino acids. No significant differences among the experimental groups were found for any of the compounds or enzymes studied. A difference approaching statistical significance was observed for amylase which was lower in the fructose and xylitol groups than in the sucrose group (Mäkinen and Scheinin, 1975C).

In view of biochemical findings in plaque and saliva during the early phases of the trial, detailed analyses of the effect of the diets were carried out on oral peroxidases, redox potential, and the concentration of ionised fluorine, iodine and thiocyanate (Mäkinen, Tenovuo and Scheinin, 1975). In the xylitol group a considerable increase in salivary lactoperoxidase activity was thus demonstrated. The concentration of fluoride was lower in the xylitol group than in the other sugar groups.

The oral biochemical studies carried out in relation to the two-year trial included an investigation of the amino-acid composition of whole saliva (Mäkinen, Lönnberg and Scheinin, 1975), and a survey of the activity of glycosidases in oral fluids and plaque (Mäkinen & Scheinin, 1975D). It was observed that the consumption of fructose, and particularly of xylitol, increased the amount of most amino acids. The increase was considered to be due to an increased protein catabolism of microorganisms deprived of sucrose. On the other hand, no increased destruction of oral collagenous tissues could be demonstrated (Mäkinen, Lönnberg and Scheinin, 1975).

No significant differences among the sugar groups were observed with regard to most glycosidases in pooled plaque or saliva preparations. However, gingival exudate showed the lowest glycosidase values in the xylitol group (Mäkinen and Scheinin, 1975D). Pooled and individual exudate samples were additionally analysed biochemically and with regard to reactions occurring in the microvasculature of the hamster cheek-pouch. These studies indicated differences in the inflammatory properties between individual exudate samples (Luostarinen *et al.*, 1975).

Observations seemingly at variance with the feeding study were obtained in the chewing-gum study with regard to lactoperoxidase and invertase-like enzyme activities; no differences were thus observed between the sucrose and xylitol groups in the latter study (Mäkinen and Scheinin, 1975D). This apparent contradiction will be discussed subsequently in this paper in view of more recent observations. On the other hand, a significant decrease in plaque formation was observed in relation to chewing *per se*. There was also a significant difference in this regard between the experimental groups. These findings were accompanied by a decrease of total colony count values during the study in both groups. No bacterial adaptation to utilise xylitol occurred during the trial (Larmas, Scheinin, Gehring and Mäkinen, 1975).

A separate study was carried out in order to examine the acute physico-chemical effects in saliva in relation to the presence of various sugars. The effects of fructose-, sucrose-, xylitol- and sorbitol-containing chewing gums were thus compared to unsweetened gum base and paraffin chewing with regard to electrolyte concentration and pH changes occurring in saliva. Stimulation of salivary secretion, due to the influence of the sugars tested, resulted in increased electrolyte concentrations with respect to Na^+, Ca^{2+} and

HCO_3^- ions in saliva. This was accompanied by a general increase in salivary pH values, notably in the case of xylitol (Söderling *et al.*, 1975).

RECENT EXPERIMENTAL, MICROBIOLOGICAL AND BIOCHEMICAL XYLITOL STUDIES

Experimental Caries

Several experimental studies have demonstrated an extremely low caries rate in relation to a xylitol-containing diet (Brin and Miller, 1974; Büttner and Moll, 1977; Gehring and Karle, 1977; Grunberg, Beskid and Brin, 1972; Karle, 1977; Karle and Gehring, 1975 and 1976; Mühlemann, Regolati and Marthaler, 1970). Total substitution of sucrose with xylitol thus resulted in caries incidence comparable with the sucrose-free controls (Karle, 1977) or still lower levels (Mühlemann *et al.*, 1977). The latter observations obtained when 25% of the diet consisted of xylitol were accompanied by pronounced diarrhoea and intestinal disturbances, which are at variance with several other experimental studies with corresponding levels of xylitol supplemented diet (Förster, Heller and Fortmeyer, 1975; Förster, 1977; Gehring and Karle, 1977; Karle, 1977). Recently, even partial substitution of dietary sucrose by xylitol, that is when both are fed together, was shown to reduce the cariogenicity of sucrose (Büttner and Moll, 1977; Green and Leach, 1977).

Microbiological Aspects

Microbial growth and metabolism in the presence of xylitol have been recently reviewed (Mäkinen, 1976). Further observations have shown that only a few strains were able to decompose xylitol at a slow rate (Ahldén *et al.*, 1977; Edwardsson, Birkhed and Mejàre, 1977; Karle and Gehring, 1976). None of the microbial species generally associated with caries metabolised xylitol, as also recently confirmed (Havenaar *et al.*, 1977). Other important observations include the absence of inoculated *S. mutans* in xylitol animals (Gehring and Karle, 1977), the lack of adaptation of *S. mutans* to metabolise xylitol despite prolonged exposures (Knuutila and

Mäkinen, 1977), and the inability of the mixed microflora to bind radioactive xylitol (Mäkinen and Rekola, 1976; Mäkinen, 1976).

Acid Formation

In accordance with the microbiological observations, no pH drop was recorded when incubating *S. mutans* and *Actinomyces viscosus* in the presence of xylitol (Noguchi and Mühlemann, 1976). This was also recorded *in vivo* using a telemetric assessment of interdental pH changes during and after xylitol rinses and chewing of xylitol-containing gum (Mühlemann *et al.*, 1977). Rinsing with a 10% xylitol solution or chewing xylitol gum subsequent to sucrose exposure was thus observed to restore the pH to the neutral level. On the other hand, reversing the order of rinsing to xylitol prior to sucrose resulted in the fast decreases of plaque pH characteristic for sucrose.

Assessment of Plaque

Short-term studies on subjects maintaining their normal dietary habits showed that chewing xylitol gum in comparison with sucrose gum reduced the amount of plaque (Mouton, Scheinin and Mäkinen, 1975A, 1975B, 1975C). Further findings in human volunteers on their habitual diet showed, however, that the colonisation of early plaque could not be inhibited by daily xylitol rinses during two days (Mühlemann *et al.*, 1977). Similarly, with unaltered diets resulting in heavy plaque formation, neither a sucrose nor a xylitol-containing chewing gum had significant effects on the plaque scores (Plüss, 1977).

Biochemical Aspects

In the present context some observations in plaque and oral fluid are mentioned due to their relation to the findings in the clinical caries trials with xylitol. Thus it was recently observed that the intake of a single dose of xylitol-containing candy induced an increased lactoperoxidase (LPO) level in human parotid saliva (Harper *et al.*, 1977). Similar observations were also made in xylitol-fed monkeys, salivary LPO activity being several times higher in the xylitol group than in the control sucrose group (Mäkinen *et al.*, 1976).

The findings in the Turku chewing-gum study (Mäkinen and

Scheinin, 1975D) in which no increase of salivary LPO was shown may thus be explained as being due to the time lapse between xylitol intake and LPO determination, the latter being predominantly carried out on morning samples.

Extracellular polysaccharide, formed through microbial *S. mutans* synthesis from sucrose, was considerably influenced by the presence of xylitol in the growth medium (Wood, 1977). Estimation of polysaccharide indicated that with increase in xylitol concentration of the medium, the production of water-soluble polysaccharide increased and the ratio of water-soluble to water-insoluble polysaccharide increased from about 0·5 in the absence of xylitol to about 4·0 in the presence of 10% xylitol. Such a development also occurring *in vivo* was indicated in studies involving an assessment of plaque adhesion in relation to chewing of xylitol-containing gum (Rekola *et al.*, 1977). The adhesivity of plaque was thus investigated through planimetric analysis of plaque, the ratio between values obtained after and before cleansing being interpreted as measurement of the adhesion (Rekola and Scheinin, 1977).

Experimental Cariogenicity

The experimental cariogenicity of several sugars, including xylitol, was tested by submitting bovine enamel surfaces covered by dacron gauze to simulated cariogenic conditions in the human mouth (Koulourides *et al.*, 1976). The decalcifying capacity of xylitol was found to equal water, and the demonstrated lack of cariogenicity of xylitol concluded to be in accord with the Turku two-year feeding study.

DISCUSSION

The weaknesses of the clinical trials of the Turku studies are recognised by the authors (pp. 26–30, 80–81, 84, 313–315; Scheinin and Mäkinen, 1975). Additional criticism has been expressed with regard to use of unconventional and sophisticated methods for scoring caries, e.g. the microscopical assessment of buccally located lesions (Von der Fehr, Löe and Theilade, 1970) and new caries indices (Scheinin, Mäkinen and Ylitalo, 1974 and 1975;

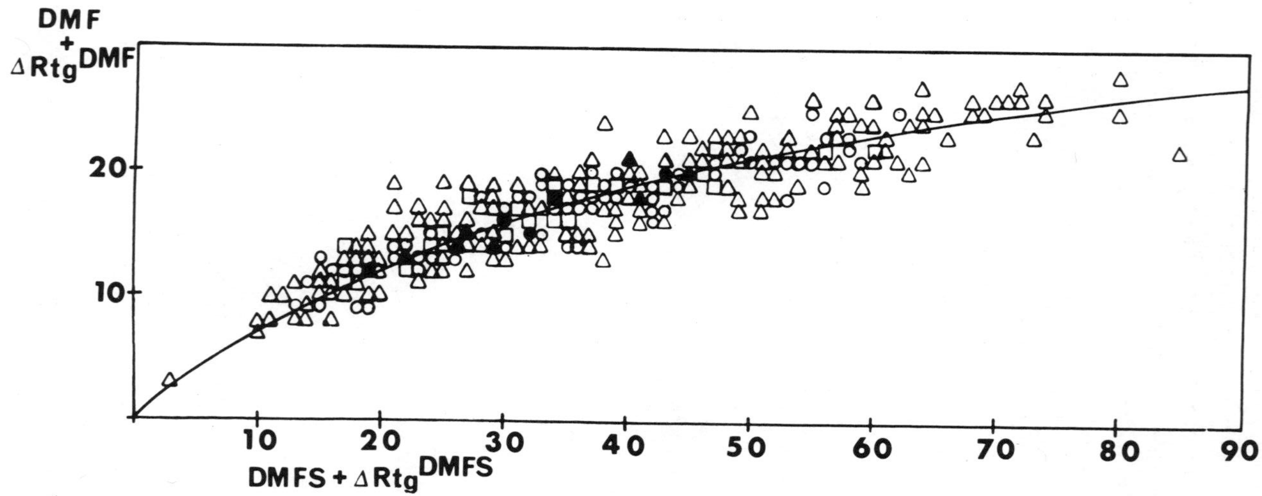

FIG. 7. Correlation between DMF and DMFS indices in 394 university students. Mean age 20 years. Mean DMF index 17·37, S.D. 4·36. Mean DMFS index 37·13, S.D. 15·06. △, 1 Observation; ○, 2 observations; □, 3 observations; ▲, 4 observations; ●, 5 observations; ■, 6 observations. (Scheinin, Honka and Kankkunen, 1970.)

Scheinin *et al.*, 1975). It should be noted, however, that the caries incidence values neither in the two-year feeding study, nor in the chewing-gum trial included the findings from the microscopic registrations. The caries incidence in the Turku studies was expressed by using the following indices:

1. The DMFS index.
2. A modified DMFS index, including in addition to primary caries also the secondary caries development (Fig. 1).
3. A Caries Activity index, incorporating in addition to the total quantitative development (=2) also the qualitative changes in terms of lesion size.

The caries increment rate as expressed by the DMFS index in the above studies is presented in conventional terms. The caries incidence using the modified DMFS and the Caries Activity indices are thought to provide additional information of that part of the caries development which remains unrecognised in the basic DMFS index. It should also be noted that additional information may be obtained through separate analyses of the carious lesions, at the C1 level, with no break in the enamel surface, or at the C2 level when there is loss of enamel and usually dentinal involvement (Fig. 1). These data are provided in the original communications (Scheinin, Mäkinen and Ylitalo, 1974 and 1975).

Further attention should also be given to the fact that the caries prevalence values in all experimental groups of the Turku studies at the commencement of the trials were already at a level rendering the use of the DMF index meaningless. This fact is recognised through examination of the DMF and DMFS indices (Fig. 7) in a university student group comparable with the subjects of the Turku studies (Scheinin, Honka and Kankkunen, 1970).

Virtually all available experimental data conform with the microbiological and biochemical observations included in the Turku studies. In experimental animals, the findings show xylitol to be non-cariogenic. In addition, at close examination of these results, indications of a therapeutic effect exist. At present, however, properly planned animal studies are required to identify the conditions for remineralisation of carious lesions due to peroral xylitol. It seems that such study conditions may be created through arranging a feeding regimen timing xylitol subsequent to administration of a sucrose-containing cariogenic diet, particularly if the carious lesions were allowed to develop without xylitol supplementation in the initial phase of the study.

The behaviour of xylitol in microbial metabolism offers considerable advantages; so far even the adaptation of microbial strains to xylitol seems unlikely to occur. Similarly, the observations with regard to pH levels in bacterial cultures *in vitro* and human dental plaque *in vivo* indicate the potential value of peroral xylitol, particularly in neutralising existing low pH levels due to the effect of fermentable sugars and acidic beverages.

Reduced formation of plaque further enhances the effects of orally administered xylitol on the CFU values of the acidogenic oral flora. So far, the microbiological development in studies reporting for a decrease in *S. mutans*, *Lactobacillus acidophilus* or the total acidogenic flora has been expressed in terms of numerical changes of the bacterial counts. Calculations of these CFU values expressed on a volume or weight unit base of plaque would yield still more favourable values in conditions influenced by xylitol.

It is concluded that peroral xylitol offers no single spectacular mechanisms to explain the low caries incidence observed in the clinical trials and experimental studies so far carried out. These findings have to be viewed as natural consequences of the microbiological and biochemical behaviour of xylitol. These include the stimulation of salivary secretion, the elevation of certain electrolyte concentrations of oral fluid, the increased buffering of saliva, the increase and/or maintenance of high pH in oral fluid and plaque, the lack of suitability for microbial metabolism, increased salivary lactoperoxidase and probably also the increased content of certain amino acids or oral fluid. The latter observations, particularly with regard to arginine, histidine, lysine and glycine (Mäkinen, Lönnberg and Scheinin, 1975) may well be related to the potential rôle of sialin, an arginine peptide, in the prevention of dental caries (Kleinberg, Kanapka and Craw, 1976). As an entity, peroral xylitol should thus be considered to stimulate a number of existing defence mechanisms against caries.

REFERENCES

Ahldén, M.-L., Birkhed, D., Edwardsson, S. & Frostell, G. (1977) Quantitative determination of the acid production from maltitol, sorbitol and xylitol by oral streptococci and lactobacilli. To be published.

Bradley, J. V. (1968) *Distribution-free Statistical Tests*. Englewood Cliffs, N.J.: Prentice-Hall.

Brin, M. & Miller, O. N. (1974) The safety of oral xylitol. In: *Sugars in Nutrition*, eds. H. L. Sipple & K. W. McNutt, pp. 590–606. New York, San Francisco & London: Academic Press.

Büttner, W. & Moll, R. (1977) Caries incidence in the rat following partial substitution of sucrose by xylitol. *ORCA Congress Abstracts*, Megeve, 1977.

Chilton, N. W. (1967) *Design and Analysis in Dental and Oral Research*. Philadelphia & Toronto: Lippincott.

COCSTOC (Commission on Classification and Statistics for Oral Conditions) (1973) Principal requirements for controlled clinical trials of caries preventive agents and procedures, eds. H. S. Horowitz, L. J. Baume, O. Backer Dirks, G. N. Davies & G. L. Slack, *Int. Dent. J.*, **23**, 506–16.

Edwardsson, S., Birkhed, D. & Mejàre, B. (1977) Acid production from maltitol in comparison with that from Lycasin®, sorbitol and xylitol by oral streptococci and lactobacilli. To be published.

Förster, H. (1977) Nebenwirkungen von Zuckeraustauschstoffen. *Dtsch. zahnärztl. Z.*, in press.

Förster, H., Heller, G. & Fortmeyer, H. P. (1975) Zur Bedeutung des Kohlenhydratanteils in einer synthetischen Diät. *Infusionstherapie*, **2,** 327.

Gehring, F. & Karle, J. E. (1977) Cariogenicity of L-sorbose compared with xylitol and sucrose in animal studies. *ORCA Congress Abstracts*, Megeve, 1977.

Gehring, F., Mäkinen, K. K., Larmas, M. & Scheinin, A. (1974) Turku sugar studies. IV. An intermediate report on the differentiation of polysaccharide-forming streptococci (*S. mutans*). *Acta Odont. Scand.*, **32,** 435–44.

Gehring, F., Mäkinen, K. K., Larmas, M. & Scheinin, A. (1975) Turku sugar studies. X. Occurrence of polysaccharide-forming streptococci and ability of the mixed plaque microbiota to ferment various carbohydrates. *Acta Odont. Scand.*, **33,** Suppl. 70, 223–37.

Green, R. M. & Leach, S. A. (1977) Effect of xylitol dietary supplements on caries in the rat. *IADR Congress Abstracts*, Copenhagen, 1977. Abstract No. 207.

Grunberg, E., Beskid, G. & Brin, M. (1972) Efficacy of xylitol in reducing dental caries in rats. *Abstr. 9th Int. Cong. Nutr.*, 1972.

Harper, L. R., Poole, A. E. & Wolf, S. I. (1977) Xylitol stimulation of lactoperoxidase in human parotid saliva. *IADR Congress Abstracts*, Copenhagen, 1977. Abstract No. 78.

Havenaar, R., Huis, in't Veld, J. H., Backer Dirks, O. & de Stoppelaar, J. D. (1977) *ORCA Congress Abstracts*, Megeve, 1977, Abstract No. 84.

Huttunen, J. K., Mäkinen, K. K. & Scheinin, A. (1975) Turku sugar studies. XI. Effects of sucrose, fructose and xylitol diets on glucose, lipid and urate metabolism. *Acta Odont. Scand.*, **33,** Suppl. 70, 239–45.

Karle, E. J. (1977) Die Kariogenität von Xylit im Tierversuch. *Dtsch. zahnärztl. Z.*, in press.

Karle, E. J. & Gehring, F. (1975) Wirkung der Zuckeraustauschstoffe Fruktose, Sorbit und Xylit auf Kariesbefall und Plaqueflora der Ratte. *Dtsch. zahnärztl. Z.*, **30,** 356.

Karle, E. J. & Gehring, F. (1976) Zur kariogenität xylitabbauender Keime im gnotobiotischen Versuch. *Dtsch. zahnärztl. Z.*, **31,** 22–5.

Kleinberg, I., Kanapka, J. A. & Craw, D. (1976) Effect of saliva and salivary factors on the metabolism of the mixed oral flora. In: *Proceedings on Microbial Aspects of Dental Caries*, eds. H. M. Stiles, W. J. Lösche & T. C. O'Brien. Sp. Suppl. *Microbiology Abstracts*, Vol. II, pp. 433–64.

Knuutila, M. L. E. & Mäkinen, K. K. (1977) Extracellular proteolytic activity of the cells of *Streptococcus mutans* grown on glucose or xylitol. To be published.

Koulourides, T., Bodden, R., Keller, S., Manson-Hing, L., Lastra, J. & Housch, T. (1976) Cariogenicity of nine sugars tested with an intraoral device in man. Experimental cariogenicity of sugars. *Caries Res.*, **10,** 427–41.

Larmas, M., Mäkinen, K. K. & Scheinin, A. (1975) Turku sugar studies. VIII. Principal microbiological findings. *Acta Odont. Scand.*, **33,** Suppl. 70, 173–216.

Larmas, M., Scheinin, A., Gehring, F. & Mäkinen, K. K. (1975) Turku sugar studies. XX. Microbiological findings and plaque index values in relation to 1 year use of xylitol chewing gum. *Acta Odont. Scand.*, **33,** Suppl. 70, 321–36.

Luostarinen, V., Paunio, K. U., Varrela, J., Rekola, M., Luoma, S., Scheinin, A. & Mäkinen, K. K. (1975) Turku sugar studies. XV. Vascular reactions in the hamster cheek pouch to human gingival exudate. *Acta. Odont. Scand.*, **33,** Suppl. 70, 287–91.

Mäkinen, K. K. (1976) Microbial growth and metabolism in plaque in the presence of sugar alcohols. In: *Proceedings on Microbial Aspects of Dental Caries*, eds. H. M. Stiles, W. J. Lösche & T. C. O'Brien. Sp. Suppl. *Microbiology Abstracts*, Vol. II, pp. 521–38.

Mäkinen, K. K. & Rekola, M. (1976) Xylitol binding in human dental plaque. *J. Dent. Res.*, **55,** 900–4.

Mäkinen, K. K. & Scheinin, A. (1975A) Turku sugar studies. VI. The administration of the trial and the control of the dietary regimen. *Acta Odont. Scand.*, **33,** Suppl. 70, 105–27.

Mäkinen, K. K. & Scheinin, A. (1975B) Turku sugar studies. VII. Principal biochemical findings on whole saliva and plaque. *Acta Odont. Scand.*, **33,** Suppl. 70, 129–71.

Mäkinen, K. K. & Scheinin, A. (1975C) Turku sugar studies. XIII. Effect of the diet on certain clinico-chemical values of serum. *Acta Odont. Scand.*, **33,** Suppl. 70, 265–76.

Mäkinen, K. K. & Scheinin, A. (1975D) Turku sugar studies. XIX. Salivary peroxidase and invertase-like activity in relation to 1 year use of sucrose and xylitol chewing gum. *Acta Odont. Scand.*, **33,** Suppl. 70, 317–20.

Mäkinen, K. K., Bowen, W. H., Dalgard, D. & Fitzgerald, G. (1976) Effect of peroral administration of xylitol on exocrine secretions of monkeys. *J. Nutr.*, in press.

Mäkinen, K. K., Lönnberg, P. & Scheinin, A. (1975) Turku sugar studies. XIV. Amino acid analysis of saliva. *Acta Odont. Scand.*, **33,** Suppl. 70, 277–86.

Mäkinen, K. K., Mielityinen, H. & Scheinin, A. (1975) Turku sugar studies. XVI. Disc electrophoresis of serum. *Acta Odont. Scand.*, **33,** Suppl. 70, 293–5.

Mäkinen, K. K., Tenovuo, J. & Scheinin, A. (1975) Turku sugar studies. XII. The effect of the diet on oral peroxidases, redox potential and the concentration of ionized fluorine, iodine and thiocyanate. *Acta Odont. Scand.*, **33,** Suppl. 70, 247–63.

Mouton, C., Scheinin, A. & Mäkinen, K. K. (1975A) Effect on plaque of a xylitol containing chewing-gum. A pilot study. *Acta Odont. Scand.*, **33,** 27–31.

Mouton, C., Scheinin, A. & Mäkinen, K. K. (1975B) Effect on plaque of a xylitol containing chewing gum. A clinical and biochemical study. *Acta Odont. Scand.*, **33,** 33–40.

Mouton, C., Scheinin, A. & Mäkinen, K. K. (1975C) Effect of a xylitol chewing gum on plaque quantity and quality. *Acta Odont. Scand.*, **33,** 251–7.

Mühlemann, H. R., Regolati, B. & Marthaler, T. M. (1970) The effect on rat fissure caries of xylitol and sorbitol. *Helv. Odont. Acta*, **14,** 48–50.

Mühlemann, H. R., Schmid, R., Noguchi, T., Imfeld, Th. & Hirsch, R. S. (1977) Some dental effects of xylitol under laboratory and *in vivo* conditions. *Caries Res.*, in press.

Nizel, A. E. (1972) *Nutrition in Preventive Dentistry: Science and Practice*, p. 356. Philadelphia: W. B. Saunders.

Noguchi, T. & Mühlemann, H. R. (1976) The effects of some carbohydrates on *in vitro* growth of *Streptococcus mutans* and *Actinomyces viscosus*. *Schweiz. Wschr. Zahnheilk.*, **86,** 1361–70.

Paunio, K., Mäkinen, K. K., Scheinin, A. & Ylitalo, K. (1975) Turku sugar studies. IX. Principal periodontal findings. *Acta Odont. Scand.*, **33,** Suppl. 70, 217–22.

Plüss, E. M. (1977) Effect on plaque growth of xylitol and sucrose containing chewing gums. *J. Clin. Per.*, in press.

Rekola, M. & Scheinin, A. (1977) Quantification of dental plaque through planimetric analysis. *Scand. J. Dent. Res.*, **85,** 51–5.

Rekola, M., Läikkö, I., Anttinen, H. & Scheinin, A. (1977) Screening of potential cariogenicity of polyol mixtures in chewing gum. To be published.

Scheinin, A. & Mäkinen, K. K. (1975) Turku sugar studies. I–XXI. *Acta Odont. Scand.*, **33,** Suppl. 70, 1–350.

Scheinin, A., Honka, K. & Kankkunen, S. (1970) Dental conditions and need for dental treatment among university students in Turku. I. Operative and endodontic treatments. *Acta Odont. Scand.*, **28,** 503–21.

Scheinin, A., Mäkinen, K. K., Tammisalo, E. & Rekola, M. (1975) Turku sugar studies. XVIII. Incidence of dental caries in relation to 1 year consumption of xylitol chewing gum. *Acta Odont. Scand.*, **33,** 269–78.

Scheinin, A., Mäkinen, K. K. & Ylitalo, K. (1974) Turku sugar studies. I. An intermediate report on the effect of sucrose, fructose and xylitol diets on the caries incidence in man. *Acta Odont. Scand.*, **32,** 383–412.

Scheinin, A., Mäkinen, K. K. & Ylitalo, K. (1975) Turku sugar studies. V. Final report on the effect of sucrose, fructose and xylitol diets on the caries incidence in man. *Acta Odont. Scand.*, **33,** Suppl. 70, 67–104.

Söderling, E., Rekola, M., Mäkinen, K. K. & Scheinin, A. (1975) Turku sugar studies. XXI. Xylitol, sorbitol, fructose and sucrose induced physico-chemical changes in saliva. *Acta Odont. Scand.*, **33,** Suppl. 70, 337–343.

Von der Fehr, F. R., Löe, H. & Theilade, E. (1970) Experimental caries in man. *Caries Res.*, **4,** 131–48.

Wood, J. M. (1977) An investigation of the effect of xylitol on the production of polysaccharide from sucrose by *Streptococcus mutans.* To be published.

CHAPTER 10

Some Aspects of Forthcoming Research on Xylitol

F. GEY
(*Xyrofin Limited, Baar, Switzerland*)

New research projects have to be built upon previously established facts because the expected results can only extend the present knowledge by cautious evaluation of new concepts, speculations and indications both with regard to the mechanism of accepted actions and/or the discovery of new actions. Considering xylitol in particular, its present position with respect to accepted facts is described in the 1977 edition of *Accepted Dental Therapeutics* (American Dental Association) as follows:

> 'Results from the chewing gum study parallelled those in the dietary replacement study and supported the conclusion that xylitol was *non-cariogenic*. The suggestion has been made that xylitol may have an *anticariogenic effect* . . . but this conclusion is not universally accepted by dental specialists. Further investigation is needed.'

The careful wording makes two important points for forthcoming research:

1. Xylitol is the first nutritive sucrose substitute which has been proved to be non-acidogenic and, even more importantly, according to a two-year clinical trial to be *non-cariogenic*. Since the non-cariogenicity of xylitol has been accepted by

the majority of experts, it is in any case a keystone for the future. Even if no other aspects were to be revealed, xylitol should nevertheless remain pre-eminent among the presently known nutritive sucrose substitutes because, in addition to its non-cariogenicity, xylitol combines other relevant properties, e.g. full sweetness (comparable to sucrose), natural occurrence in the common human diet and in the normal mammalian metabolism.

2. In the one-year chewing-gum trial in Turku a comprehensive substitution of between-meal sucrose by xylitol-containing chewing gum prevented the increment of caries to the same degree as an almost complete sucrose substitution. However, since even far-reaching substitution of sucrose between meals means only a partial substitution of the total dietary sucrose, *some (presumably indirect) anticariogenic* properties have been suggested by the authors of the Turku studies (Scheinin and Mäkinen, 1975). Even the mere possibility of any anticariogenic potential makes xylitol of the utmost interest for various public health organisations and requires to be clinically evaluated for these potentials. At present, several institutions are planning to re-investigate xylitol-containing chewing gum. For instance, the National Institute for Dental Research/National Institute of Health of the USA intends to carry out *clinical trials* with hundreds of children for 2–3 years. The World Health Organisation considers *field demonstration programmes (field studies)* with chewing gum in even larger groups of children. Sooner or later, chewing-gum manufacturers may be expected to carry out additional trials with special products. Since chewing gum is not a major item of confectionery, clinical studies on other confections deserve attention and might become desirable as well, e.g. trials of xylitol candy or field studies with a variety of xylitol products which might themselves be developed into a prophylactic programme. Unfortunately, it will be at least several years before the data from any such comprehensive trials are forthcoming for definitive evaluation of the xylitol potential.

As long as it will be some time before the results of the additional clinical trials are available, some support may be expected from *short-term experiments* on cariogenic bacteria, experimental animals, human plaque, mineral dynamics of enamel, etc. They may not only provide *circumstantial evidence* but may, in addition, elucidate possible mechanisms of xylitol action. A few examples of conceivable lines of promising research are outlined below.

MICROBIOLOGY

It has been well established that the cariogenic bacteria of dental plaque, e.g. *Streptococcus mutans*, use glucose or sucrose to the same extent in the presence of xylitol as in its absence, i.e. that xylitol has no direct bacteriostatic action. Nevertheless, this does not exclude other interference of xylitol with the metabolism of cariogenic fuels like sucrose. Wood tested in a pilot experiment (Table 1) whether xylitol affects the ratio of insoluble/soluble polysaccharides (Wood, 1976). According to his preliminary data, xylitol (in concentrations as present in the oral cavity during consumption of a xylitol chewing gum) seems indeed to favour the formation of *water-soluble polysaccharides* instead of water-insoluble polysaccharides: the ratio of water soluble/water insoluble polysaccharides was at least trebled by xylitol. At the same time, there were also gross differences

TABLE 1

Effect of Xylitol on the Production of Extracellular Polysaccharide from Sucrose by *Streptococcus mutans*[a]

Culture		*Polysaccharide produced from sucrose*			
Xylitol (%w/v)	*Final pH*	*Watersol. (μg/ml broth)*	*Waterinsol. (μg/ml broth)*	*Total*	*Ratio watersol. to waterinsol.*
0	4·0	24	44	68	0·54
0·1	4·0	35	26	61	1·34
2·0	4·0	44	29	73	1·51
10·0	5·8	85	22	107	3·86

[a] Brain/heart infusion broth with 10% (w/v) sucrose (48 h; 37 °C). (Wood, 1976.)

in ethanol precipitates of water-soluble polysaccharides as investigated by electron microscopy: media not containing xylitol produced water-soluble polysaccharides with fibrous structures whereas the presence of 2% xylitol resulted in an amorphous structure and 10% xylitol in a granular structure. Since water solubility of polysaccharides may be of crucial importance for the adherence and colonisation of dental plaque, these xylitol effects could be, at least in part, related to a conceivably decreased plaque formation (see below)—provided that xylitol has the same effect in plaque *in vivo*. Bacterial adhesion, being a central phenomenon in the pathogenesis of plaque, should be investigated in the future with a variety of techniques ideally including *in vivo* assessments.

CARIES IN EXPERIMENTAL ANIMALS

In several conventional caries experiments in weanling rats, continuous substitution of dietary sucrose by xylitol resulted in reduction of caries rates which even seemed to be better than in the case of sorbitol (Mühlemann *et al.*, 1970; Karle and Büttner, 1971; Grunberg *et al.*, 1971; Karle and Gehring, 1975). The interpretation of these results has, however, been hampered by two facts.

1. Rats not adapted after weaning to rather high dietary levels of sucrose substitutes are particularly susceptible to the polyols sorbitol, xylitol and mannitol, and even to sucrose (Mühlemann *et al.* 1970; Grunberg *et al.*, 1971; Hosoya and Iitoyo, 1969A; Hosoya and Iitoyo, 1969B; Mühlemann *et al.*, 1977). Weanling rats, being more sensitive than young or adult rats, are known to show strong initial diarrhoea (with all secondary disturbances) and may even survive only in part. This temporary sensitivity involves mainly a logistic problem for the experimental caries since rats develop sucrose-induced caries most rapidly after weaning, but laxation affects the eating and drinking habits of the rats and thus their caries increment (König, 1971). In consequence, in

the future only those protocols should be used in animal experiments which prevent major and/or consistent laxation and changes in drinking habits.

2. Since solid sucrose between meals is the most cariogenic fraction of the diet of humans, a constant ratio of combined xylitol and cariogenic sucrose in a rat diet would not reflect the 'normal' life of humans.

One method of preventing persistent laxation is to *raise the xylitol dosage stepwise* to 20% of the diet with successive increases of 5% per week. Even weanling rats have been shown to tolerate 5% xylitol (Hosoya and Iitoyo, 1969A). This adaptation procedure was demonstrated to avoid both persistent laxation and reduced weight gain of growing rats (Hosoya and Iitoyo, 1969A). At this slow increase of the xylitol dosage however, it will not be possible to measure alterations of the caries increment in the conventional three-week period after weaning. In consequence, more prolonged periods for the development of carious lesions have to be studied in the future, if xylitol and sucrose are simultaneously present in the diet.

Another method of partially reducing the laxative effect of xylitol as well as mimicking the intermediate consumption of xylitol by humans involves *frequent topical application* between meals. The following protocol was designed in order to simulate 'snacks' between meals (Kinkel and Gey, 1977): male Wistar rats after weaning received the same restricted amount (total 5 g—later 6 g—daily) of the cariogenic sucrose-containing Standard Diet 2000 (with 56% sucrose), fed in 3 portions at fixed times. Twice after each feeding period a very small volume of the sucrose substitutes was topically applied, i.e. the 60% solution of the sucrose substitutes was dropped onto the teeth and painted over them using a soft artist's brush. These 'snacks' representing 10% of the diet were given 30 and 90 min after each feeding period. After 5 weeks, the number of molar lesions showed the following order: sucrose increased the number of lesions significantly in comparison to both controls, i.e. by about 35% as compared with controls treated with water and by about 50% in comparison with controls without any treatment. Sorbitol and sorbose had neither a positive nor a negative effect since the values of sorbitol- and sorbose-treated rats were in the range of both control groups. Xylitol was the only compound that diminished the number of carious lesions significantly in comparison with both control groups (by about 20 % and 27 % in comparison with water treated and untreated controls, respectively) (Table 2).

TABLE 2

Effect on Experimental Rat Caries (molar lesions, state B) of Frequent, Intermittent Topical Administration of Xylitol and Other Sucrose Substitutes[a] Between Restricted Standardised Cariogenic Meals (Standard Diet 2000 with 56 % Sucrose). (After Kinkel and Gey, 1977)

Painted 'snacks' of substit.		*Molar lesions (dentino-enamel junction)*	*Final body weight[b]*
Sucrose	($n = 45$)	$8{\cdot}0 \pm 0{\cdot}2^d$	66 ± 1
None	($n = 45$)	$5{\cdot}9 \pm 0{\cdot}2^c$	71 ± 1
Sorbitol	($n = 25$)	$5{\cdot}9 \pm 0{\cdot}3^c$	58 ± 2^d
Water	($n = 25$)	$5{\cdot}4 \pm 0{\cdot}2$	68 ± 2
Sorbose	($n = 20$)	$5{\cdot}0 \pm 0{\cdot}4^c$	65 ± 2
Xylitol	($n = 44$)	$4{\cdot}3 \pm 0{\cdot}3^d$	60 ± 1^d

[a] Twice after each of three meals (substitutes total 10 % of diet as 60 % solution).
[b] Initial body weight, 32 ± 1 g; duration of the experiment, 5 weeks.
[c] $p > 0{\cdot}05$ (in comparison to water-treated controls).
[d] $p < 0{\cdot}01$ (in comparison to water-treated controls).

The restricted feeding pattern reduced the growth in all animal groups and also substantially diminished the laxation effect. But still the final body weight in the xylitol-treated rats was 12 % lower than that of water-treated controls, the weight of sorbitol treated animals was 15 % lower, and thus comparably low (Table 2). Although the faeces of the polyol-treated rats showed some initial softness, the intestinal disturbances by xylitol may hardly be a major factor for the reduction of carious lesions since the sorbitol-treated rats showed at least the same reduction in weight gain whereas sorbitol lacked any significant beneficial effect on the carious lesions. This type of protocol with topical application of sucrose substitutes, will, of course, be repeated with additional stepwise adaptation of the weanling rats to xylitol as mentioned above.

Another type of intermittent xylitol application in weanling rats was recently investigated in the UK by Green and Leach who tried to simulate the intermediate consumption of xylitol as well (Green and Leach, 1977). The animals received every second day the cariogenic

TABLE 3

Effect of Alternate-day Feeding of a Xylitol–Starch Diet and a Cariogenic Sucrose Diet on Experimental Weanling Rat ($n = 16$) Caries. (After Green and Leach, 1977.)

Alternate-day feeding pattern *Even days*	*Odd days*	*Total molar lesions*	*Very advanced lesions*	*Weight gains (g)*
66 % sucrose	66 % starch only	3·2 ± 2·3	0·6	70 ± 12
66 % sucrose	60 % starch + 6 % xylitol	1·2 ± 1·4[a]	0·1	56 ± 8[a]

[a] $p < 0{\cdot}01$ in comparison to control rats fed starch only.

standard diet with sucrose, alternating with a sucrose-free starch diet either with or without xylitol. This *alternate-day feeding* of xylitol significantly reduced the number of molar lesions by 62 % in comparison with the controls receiving starch without xylitol on the alternate day (Table 3). Unfortunately, in this new type of protocol also, the gain in body weight was still 20 % lower than in controls. Here, again, it is desirable, by stepwise adaptation, to eliminate the difference in body weight.

The above preliminary studies in rats suggest that, at least under these experimental conditions with *repeated consumption of xylitol subsequent to regular cariogenic meals*, xylitol is superior to both sorbitol and sorbose since xylitol can reduce the caries incidence significantly beyond that of controls (Tables 2 and 3). Therefore, additional rat experiments with these or, hopefully, even improved protocols, might come to agreement with some conceivable anticariogenic potentials of xylitol.

PHYSICO-CHEMICAL PROPERTIES OF SALIVA AS STIMULATED BY XYLITOL

The authors of the Turku studies discussed the possibility that the salivary response to oral xylitol (including, for example, an increase in salivary flow, pH and calcium content) may be of crucial importance for the result of their chewing-gum study (Scheinin and Mäkinen, 1975). Actually, changes in the same direction are generally considered beneficial with regard to the well-known, physiological, remineralising properties of normal saliva. In consequence, two routes of research may deserve particular interest for the future. (1) Studies on the particular *remineralising properties of xylitol-induced saliva*, by using enamel slabs by *in vitro* techniques (Yaari and Bibby, 1976) or *in vivo* procedures (Koulourides *et al.*, 1976). (2) Studies on *xylitol vehicles other than chewing gum*. Indeed, the comparison of chewing gum and hard-boiled xylitol candies (Bühler and Gey, 1977) indicates that different xylitol vehicles may show only quantitative differences regarding the increase of salivary flow, pH and calcium content (Fig. 1). This may be important with regard to the fact that sucking candies results in markedly less mastication (and thus less mechanical stimulation of salivation) than does chewing gum.

The pH curves of Fig. 1 demonstrate that the pure sucrose candy does not increase the pH significantly but rather tends to decrease the pH of total saliva. This failure of sucrose to increase the pH is masked in the sucrose chewing gum for a short, initial period by the chewing effect which is shown by chewing the gum base (Fig. 1). In consequence, any conceivable, pH-dependent benefits of xylitol (as for example, remineralisation of 'white spots') might become more obvious by the comparison of sucrose and xylitol in candy than in chewing gum. This will certainly be even more true if sites are taken into account where sucrose will be rapidly metabolised to lactic acid as in plaque (for details see Frostell, 1977, and Scheinin, 1977). Studies with xylitol candy may thus hopefully be expected to elucidate the question of whether the clinical effect of the xylitol-containing chewing gum in Turku was primarily due to xylitol as the major ingredient or to mastication.

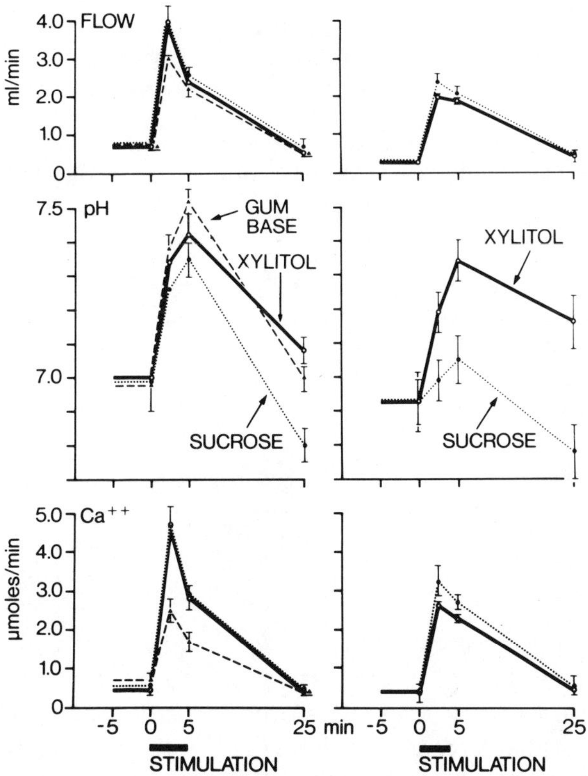

FIG. 1. Salivary responses to (left) chewing gum (2·0 g with 0% and 60% sweetener) and (right) hard-boiled candy (1·25 g without flavour). (After Bühler and Gey, 1977.)

DENTAL PLAQUE

Since the expansion of caries (as well as of periodontal diseases) depends on plaque, the effect of a short-term consumption of xylitol on plaque is of great interest. This is also true for cosmetic aims. Up to now, two short-term plaque trials in humans (Mouton, Scheinin and Mäkinen, 1975; Plüss, 1977) have shown that xylitol chewing gums cause significantly less plaque than sucrose gums. Undoubtedly, this result is of great practical importance. At the same time, it is of great scientific interest to know whether the xylitol

TABLE 4

Comparison of Two Studies on the Effect on Male Plaque Weight of Chewing Xylitol Chewing Gum for 3 Days Without Oral Hygiene

Parameter	*Mouton, Scheinin and Mäkinen*, 1975: *'Normal' diet* (*sucrose-poor?*)			*Plüss*, 1977: *Standardised diet sucrose-free*		
n	28	32	36	18	18	18
Chewing gum 5–6 pieces daily	Control (no gum)	sucrose	xylitol	gum base	sucrose	xylitol
Plaque:	Fresh weight			Dry weight		
mg	33·4	42·2	25·2	1·90	2·64	1·68
%	100	127	76	100	139	88
Significance[a] (control vs sucrose; sucrose vs xylitol)	n.s.	$p < 0.05$		$p < 0.01$	$p < 0.01$	
Significance[a] (control vs xylitol)	n.s			n.s		

[a] n.s., not significant.

gum is able to decrease plaque more markedly than in controls. Indeed, in both trials, some trend was found for the chewing of xylitol gum to reduce plaque more than either no chewing or the chewing of unsweetened gum base (Table 4); unfortunately, neither of these studies showed statistical significance. In consequence, further studies will hopefully clarify whether xylitol chewing gum or xylitol candy have any specific anti-plaque effect.

In conclusion, one should not extrapolate from any of the above-mentioned preliminary experiments in animals or men to a final classification of xylitol regarding caries. The latter remains to be elucidated in forthcoming long-term clinical trials or field demonstration programmes. Nevertheless, circumstantial evidence currently available still favours the possibility that forthcoming research might clarify additional beneficial potential of xylitol in dental caries, presently not yet fully elucidated. No one firm or no one institute may presume to identify and pursue all the promising potential of xylitol. This fact presents a challenge and an opportunity for all who are concerned with the improvement of caries prevention by sucrose substitutes.

REFERENCES

American Dental Association (1977) Council on Dental Therapeutics of the ADA. *Accepted Dental Therapeutics*, 37th edition.

Bühler, E. & Gey, F. (1977) Salivary response to xylitol chewing and xylitol candy. Unpublished results.

Frostell, G. (1977) Behaviour of xylitol in isolated plaque. This volume, p. 133.

Green, R. M. & Leach, S. A. (1977) Effect of xylitol dietary supplements on caries in the rat. *J. Dent. Res.*, **56,** A 207.

Grunberg, E., Beskid, G. & Brin, M. (1971) Xylitol and dental caries. Efficacy of xylitol in reducing dental caries in rats. *Int. J. Vit. Nutr. Res.*, **43,** 227–32.

Hosoya, N. & Iitoyo, N. (1969A) Effects of xylitol administration on the development (growth) of rats. *J. Jap. Soc. Food Nutr.*, **22**(2).

Hosoya, N. & Iitoyo, N. (1969B) Inducing effect of xylitol in rats. In: *Pentoses and Pentitols*, eds. B. L. Horecker, K. Lang & Y. Takagi, p. 197–200. Berlin–Heidelberg–New York: Springer-Verlag.

Karle, E. & Büttner, W. (1971) Kariesbefall im Tierversuch nach Verabreichung von Sorbit, Xylit, Lycasin und Calciumsaccharose-phosphat. *Dtsch. zahnärztl. Z.*, **26,** 1097.

Karle, E. & Gehring, F. (1975) Wirkung der Zuckeraustauschstoffe Fructose, Sorbit und Xylit auf Kariesbefall und Plaqueflora der Ratte. *Dtsch. zahnärztl. Z.*, **30,** 356–63.

Kinkel, H. J. & Gey, F. (1977) Effect of frequent topical application of sucrose substitutes on experimental rat caries. Unpublished results.

König, K. G. (1971) Zur Kariogenität von Nahrungszuckern und von Zuckeraustauschstoffen im Tierversuch unter Berücksichtigung unspezifischer Effekte. *Dtsch. zahnärztl. Z.*, **26,** 1093–96.

Koulourides, T., Bodden, R., Keller, S., Manson-Hing, L., Lastra, L. & Housch, T. (1976) Cariogenicity of nine sugars tested in an intra-oral device in man. *Caries Res.*, **10,** 427–41.

Mouton, C., Scheinin, A. & Mäkinen, K. K. (1975) Effect of xylitol chewing gum on plaque quantity and quality. *Acta Odont. Scand.*, **33,** 251–7.

Mühlemann, H. R., Regolati, B. & Marthaler, T. (1970) The effect on rat fissure caries of xylitol and sorbitol. *Helv. Odont. Acta*, **14,** 48–50.

Mühlemann, H. R., Schmid, R., Noouchi, T., Imfeld, Th. & Hirsch, R. S. (1977) Some dental effects of xylitol under laboratory and *in vivo* conditions. *Caries Res.*, in press.

Plüss, E. M. (1977) Effect on plaque growth of xylitol and sucrose containing chewing gums. *J. Clin. Periodontology*, in press.

Scheinin, A. (1977) Incidence of clinical and experimental caries in relation to microbiological and biochemical xylitol-induced effects. This volume, p. 139.

Scheinin, A. & Mäkinen, K. K. (1975) Turku sugar studies. I–XXI. *Acta Odont. Scand.*, **33,** Suppl. 70, 1–350.

Wood, J. M. (1976) An investigation of the effect of xylitol on the production of polysaccharides from sucrose by *Streptococcus mutans*. Report on Special Project P506, 1–9.

Yaari, A. & Bibby, B. G. (1976) Production of plaques and initiation of caries *in vitro*. *J. Dent. Res.*, **55,** 30–6.

Discussion: Dental Aspects

Chairman—Mr M. Midda

D. A. M. Geddes (*Glasgow Dental Hospital & School, Glasgow, UK*): May I ask two questions please? Firstly, have any organisms from other habitats, for instance the soil, been shown to metabolise xylitol? Secondly, do rats who show a poor weight gain also have delayed eruption of the teeth?

F. Gehring (*Bayerische Julius-Maximillians-University, Würzburg, W. Germany*): We have not found bacteria from the soil which can metabolise xylitol.

F. Gey (*Xyrofin Limited, Baar, Switzerland*): I would rather like to turn over the second question to someone more experienced in this respect. As far as I know there is not a very marked difference in the eruption of teeth. If you have a laxative effect, there is a possible retardation of body growth but I should like to get a more competent answer.

R. M. Green (*University of Wales, Dental School, Cardiff, UK*): The only teeth that we considered in our studies were erupted before the animals were fed xylitol so no effect on eruption could be evaluated.

B. Guggenhelm (*The Battelle Institute, Zurich, Switzerland*): In our experiments with xylitol, we saw a change in drinking frequencies in that the animals drink more frequently.

Gey: Yes, I fully agree. In the future you have to avoid changes on both eating and drinking habits otherwise you will never get reliable results.

S. Leach (*University of Liverpool, School of Dental Surgery, Liverpool, UK*)*:* I would like to ask Professor Gehring about the streptococci that fermented xylitol. You gave them all a code number: can you tell us which kinds of streptococci they were, for example—*sanguis*, *mutans* or *salivarius*?

Gehring: Xylitol-degrading capabilities were found only in enterococci-like bacteria, *not* in *Streptococcus mutans*, not in *Streptococcus sanguis* and not in *Streptococcus salivarius*. Not from any other streptococci, only from enterococci-like bacteria. In the plaque material of humans and in the oral cavity of rats we have found such enterococcal organisms.

Leach: If I may make a comment please. I think I would have to be very careful about the extra amount of soluble extracellular polysaccharides produced by increasing the xylitol because sucrose does precisely the same thing. High concentrations of sucrose are certainly more cariogenic than low concentrations of sucrose or so we think. Also the soluble polysaccharides are capable of being fermented in contrast to the insoluble material which, once it is formed, is inert.

Gey: Well, I fully agree that the insoluble polysaccharides are the longest lasting store of energy. I have carefully avoided extrapolation of the data of Dr Wood referring to the oral cavity *in vivo*. I think I made it quite clear that this is a very interesting finding. Professor Scheinin said something very similar in that this is a finding that remains to be re-evaluated in the human oral cavity.

T. Grenby (*Guy's Hospital, London, UK*)*:* So, could I ask what the consensus of opinion of the speakers is about the evidence available so far. Do you consider that the evidence shows that xylitol is, in the main, inert with respect to caries, neither promoting nor retarding it actively, or if you describe xylitol as anti-cariogenic, is it actively able to combat caries in some way?

A. Scheinin (*University of Turku, Turku, Finland*): I do not like to make a fool of myself or the panel so I will answer by stating my own opinion. It is difficult enough to measure reduction of caries properly, and when it comes to negative incidence (lower values than zero) I am not aware of any sound methods which could be applied clinically or radiographically because even the assessment through densitometric analysis of radiographs would include variables like projection voltage changes which would be extremely difficult to cover in a longitudinal study. However, I think that there are indications in these studies. I also think that there are observations not only in our clinical trials but also in animal experiments indicating that there is some sort of an effect which would provide an additional anti-cariogenic effect. Further studies should be carried out and I think that, as the lifespan of man is relatively short, it would be much better to try to develop a mechanism. It does not matter if this is through a biochemical approach or an animal study but the planning should include the possibility of measuring remineralisation.

M. Midda (*School for Dental Hygiene, University of Bristol, UK*): In the planometric studies that you were doing on plaque you felt that it was very difficult to evaluate the growth and also the maturity of plaques. Have you considered the use of two-tone dyes that would show not only the difference in the film thickness but also the age of the colony?

Scheinin: Yes, we have considered it, although it has not yet been used in our studies. I was not really referring to the difficulties of carrying out the planometric—that is a fairly easy and objective job—the difficulty is assessing plaque through the existing plaque indices. It does not matter whether this is a modified Quigley and Hine index or the Löe and Silness index—all these indices are excellent for showing condition with no plaque at all or some level of plaque. If one has to evaluate the efficiency of properly conducted tooth brushing then those indices are quite sufficient. The problem comes when evaluating incidences between values for sugars or sugar substitutes in conditions involving no tooth brushing during a relatively short experimental period.

Squadron Leader Bashaarat (*Institute of Dental Training, RAF Halton, Wendover, UK*): I noticed in the exhibition three toothpaste products containing xylitol; two also contain fluoride and one, Italian I believe, has chlorophyll. Would the panel care to comment on the use of xylitol in this medium.

Scheinin: To my knowledge there are xylitol toothpastes in a few countries—Finland is one, Italy is another and I have heard that some are shortly to be introduced into the United States. One has to view the situation with regard to existing composition of toothpastes. There are toothpastes with a sorbitol content of 60% and even slightly higher and little is stated on the labels of the toothpastes or packages about the sorbitol content. I do not think that it is fair to argue about the relative merits of xylitol and sorbitol in toothpaste and to try to compare them but I think that in principle a toothpaste should contain ingredients that are not metabolised by cariogenic bacteria. I also think that there might be certain measurable properties which could suggest a benefit if the polyols were included in toothpastes.

Guggenhelm: Does xylitol prevent dental disease?

Scheinin: There was a short paper included in the Turku sugar study dealing with the dental effects but the problem was that we did not measure the predominant conditions except plaque at the beginning of the study and our study also involved subjects with a relatively high level of oral hygiene. I would, however, say, in view of the lactoperoxidase findings, that there are certain indications which could recommend xylitol or polyols in general but I hesitate to say if these are of any practical value.

As a whole, a fermentable sugar would act as a source for inflammation mediators. I am referring to a paper by Brown published in 1975 and there one has not only to consider acid production and polysaccharides but also natural molecules involving the number of enzymes which could act as mediators of inflammation. From this viewpoint, sucrose would influence gingival inflammation in a negative way and this is actually true. Most clinicians know that if you remove a sucrose plaque, it is almost impossible to do so without causing bleeding at the gingival margin.

M. S. Jeffery (*Cadbury Ltd, Birmingham, UK*): In the exhibition we have seen several products on sale in various countries. I wonder whether you could say how they are being promoted and what sort of support they are getting from the dental profession in those particular countries and do you feel that you get a similar support in this country at the moment?

Gey: Well, with regard to the American situation I have read that the American Dental Association stated in their Accepted Dental Therapeutics that xylitol products were actually encouraged. There are countries like Switzerland, for instance, where you have definite classification of compounds; there xylitol has been classified as non-cariogenic based on the Turku trials, whereas other compounds like sorbitol or Lycasin are just considered as friendly to the teeth. With regard to Sweden, I would like to turn the question to either Professor Frostell or Dr Larje.

G. Frostell (*Lund University, Malmö, Sweden*): The Swedish authorities favour sucrose substitutes in between-meal candy. In Sweden we consume many pastilles and chewing gum between meals and these are considered to be risk products. The general result is now that over 50 % of chewing gum is produced with substitutes for sucrose of some kind.

O. D. Larje (*Dental Public Health Centre, Solna, Sweden*): I also would like to comment a bit about conditions in Sweden. I am vice-president of the Swedish Dental Health Association which has more than 5000 members. We are going to promote sugar substitution in our Association. One of the manufacturers which was selling one kind of sucrose/sugar-free candy talked to the Swedish Government and to the Minister of Finance and asked for a tax reduction or exemption for this confectionery because it is very expensive—as most things in Sweden! The Minister of Finance has asked seven institutions in Sweden, among them the Swedish Dental Association, to consider this matter and comment on it.

We would like to promote and recommend that there should be a tax exemption for the sugar alcohols which have been proven in scientific experiment to be less injurious to the teeth.

Midda: Of course in this country [the UK] VAT was taken off sweets and a tax put on dentistry by raising the NHS charges.

Scheinin: Just a brief comment. In my country [Finland] there will be a discussion on May 24th and 25th [1977] between the Director General of the Finnish Dental Association, a few advisers and the State Medical Board on the sugar substitutes in general.

In addition, there is one area which has not been touched upon at all during this discussion and that is those individuals with a reduced salivary secretion. This involves particular problems in treatment, generally, apart from the dental aspect. There could also be dehydrated patients as well as the dying patient and hospitals have been desperately looking for some sort of an artificial saliva. This has been touched upon in a booklet published in England in 1973 where there was an approach made to produce an artificial saliva using sorbitol and I think some paraffin, some electrolytes and carboxymethylcellulose and, if this could be used in suitable dispensers, it would take care of a lot of problems in dehydrated patients, rheumatoid arthritis cases, Sjorgren's syndrome and related conditions.

Question: There would seem to be a consensus coming out in the dental profession or is it too early?

Midda: Speaking for myself, I think it is early because xylitol is not yet available to the British public. However, there are British dentists in the audience—would they care to comment upon their interpretation of the Turku sugar studies.

Question: We have heard about the limitation of dosages in the order of 10–20 g per day. I wonder whether the manufacturers of confectionery in Finland are concerned about that, particularly when a lot of confectionery is eaten by children and whether this 10–20 g is realistic.

Scheinin: Dental doses of 20 g have been mentioned but if you look properly at the papers presented by Professor Förster and some others you will see that much higher dosages have been mentioned. Safe dosages of 30 g have been given to children as well as to diabetics. Professor Förster perhaps will comment on this: from the viewpoint of chewing gum and also hard candy 10 g is quite realistic. I think 20–30 g would also be quite realistic.

Midda: I believe the questioner was more concerned about the sort of child that perhaps eats 16 oz of candy a week.

Comment: That is right, and they do eat spectacular amounts!

H. Förster (Klinikum der Johann Wolfgang Goethe University, Frankfurt-am-Main, W. Germany): I think that there is a misunderstanding. The quantity of 10–20 g is a single dose—which can be administered 3–4 times a day. The safe dose for xylitol is 10–20 g for children—there is no diarrhoea with this single dosage. The higher dosages are not yet tested but my belief is they will be safe. Adults can take up to 30 g of xylitol as a single dose without diarrhoea. Up to 200 g or even 300 g per day have been used by adults provided that this amount was divided into single doses of 20–30 g each. The oral tolerance of xylitol is very good when it is used as a candy or a chewing gum. It is not so good when it is used in the form of soft drinks. If one litre of a 10% solution is drunk, then there might be diarrhoea. There might not be diarrhoea when only 200 cc are drunk but I think that xylitol should be used in candies or sweets rather than in drinks because the diarrhoea risk is less. It is not usual for a child to eat 100 g or more of candies within one hour or so.

Midda: It is with British children! Is there any objection to using xylitol in bakery products in these dosages—such as in cakes or pastries, or sweetened cream.

Förster: There is a greater tolerance for xylitol when it is administered with other substances in meals, particularly when only making up 10, 20 or 30% of the whole meal. The tolerance is very much higher and up to 50 g of polyols can be tolerated when included in a meal, whereas if 20–30 g of sorbitol are given as a single dose in water, then diarrhoea would be caused but I must add that the single dose and the whole dose for a day are different things. There should be no misunderstandings. When a single dose is given the single dose should not exceed 30 g, since the experiments until now have shown that 30 g is a safe dose. On the other hand, the daily intake might be 40–60 g which is still safe and does not cause diarrhoea. Some people can tolerate very much higher doses.

Midda: Thank you. Of course it is an interesting thing and I am sure somebody from the sugar industry will correct me but the pattern of sugar consumption in this country is now changing. People are turning towards more cocoa products as between meal snacks as compared to the consumption of hard candy. This may mean that the correct use of xylitol products might well be within this area because it is the between meal snacks that are certainly related to caries.

K. K. Mäkinen (*University of Turku, Turku, Finland*)*:* I have just a comment referring to the bakery products. We had a special bakery in Turku which during two years produced all the pastry we needed for the diarrhoea study and the whole assortment comprised some 30–40 different items including our national puddings, and we never found any pastry-dependent diarrhoea or flatulence during the two-year study.

Frostell: In Sweden our group had made certain recommendations as to the use of xylitol and we had suggested that since the price of xylitol is still comparatively high its use should be concentrated on so-called risk products. In our country, a certain type of pastille, sold in small boxes, is very popular. Thousands of tons of these are consumed and they are considered to be one of the main causes of the high caries frequency in Sweden. We have suggested to industry that this type of product should switch over to polyalcohols because in repeated small doses xylitol and other polyalcohols would do the best job.

I. M. Billcliffe (*Smith Kendon Ltd, Bridgend, UK*)*:* I would like to ask a question of Professor Frostell about a competitive product to xylitol—Lycasin—which has been mentioned on a number of occasions this afternoon. The last comment was that Lycasin is only 'friendly' towards teeth. My understanding of the situation is that there is definitive evidence that the latest type of Lycasin available has been shown to be as specifically non-cariogenic as xylitol. Would you care to comment on that.

Frostell: I am not aware of any study which would have shown that Lycasin of any kind should be as non-cariogenic as xylitol. On the contrary, in all studies using sorbitol, Lycasin and xylitol, the cariogenicity decreases in the order sorbitol to Lycasin to xylitol, as was shown on a slide of one of the speakers here today.

Midda: Do you know of any studies, Mr Billcliffe?

Billcliffe: I understand there was a paper, I think from Sweden, which quoted a specific type of candy manufactured with Lycasin which was shown to be non-cariogenic.

Förster: I would like to make a short comment on Lycasin. Lycasin is to a great part maltitol and tolerance to maltitol is very low. Maltitol causes diarrhoea in dosages where xylitol would be well tolerated and it has a very long lasting effect.

My volunteers said that maltitol did the most terrible things to them and they would not take part in any more tests after the first one.

J. F. Doughty (Merton House, Bootle, Liverpool, UK): One of the speakers asked for the reaction of dentists. I can only speak for a very small group of dentists. We looked at the Turku studies in some detail and were a little concerned, for example, about some of the findings and the selection of the samples. Most of these matters of concern have been answered today by the speakers. We would, however, like to see some classical epidemiological evidence of the typical studies and the results of those.

Scheinin: The typical classical index for evaluating caries in a young adult population would be the DMFS index and this was exactly the index used in the Turku study. Then there were additional indices and these I have pointed out. If you would like to assess the development in terms of dentinal lesions, that would be perfectly possible just by using a simple adding machine to combine the information given in a few tables.

Gey: May I say just one word on the terminology of non-cariogenicity? It should be in accordance with the opinion of dental authorities and be restricted to those compounds that have been proved to be non-cariogenic in a clinical trial of at least two years. Thus, non-acidogenicity in plaque trials as shown by Professor Frostell would, in the strict sense, not imply the words non-cariogenic if this effect had not been proved again in the clinical trial. Do you agree?

Frostell: Oh, yes. I do not think that you can make any statements at all on the question of whether a substance is low-cariogenic or non-cariogenic if you have not tested it carefully both in animal experiments and finally in experiments with humans.

Midda: I think we have reached the stage of knowledge that if something is non-acidogenic we can assume that it is likely to be non-cariogenic as well.

Frostell: Yes we could, it is likely to be non-cariogenic if it is completely non-acidogenic.

Guggenhelm: I think Professor Gey put the point well when he said that xylitol is the only substance which has been tested for two years and is really non-cariogenic. No-one is putting that in doubt but he should have included Lycasin and sorbitol in a similar study to see whether xylitol has a real advantage. As there is only the one Turku study I think any fair comparison in this field is at the moment not possible. Would you agree with that, Professor Scheinin?

Midda: There are certain differences in terms of consumer acceptance. Taste alone puts xylitol out ahead of the other two compounds that you have already mentioned.

Scheinin: There are studies where Lycasin and also sorbitol have been tested clinically. With regard to sorbitol there is the Meller and Postsind study carried out on school children in Denmark but there they had difficulties with the control group. There were difficulties in the Lycasin study carried out by Professor Frostell and his co-workers, for example, during the second year of the study the subjects reduced their consumption of his products.

Frostell: Yes, that is correct and in both this Lycasin study and in the maltitol double-blind study, there was not total substitution of sucrose with Lycasin, only in risk products such as candies. We found about 25–30% reduction in caries in the test groups but, statistically, this was a probable but not significant difference. However, I agree that the need for comparative studies with different sugar substitutes is urgent. We have a need since the price difference is so large that it would be of great interest to make comparative studies in humans between different sugar alcohols and sugar alcohol substitutes but, as you know, this alternative is expensive to carry out. In the meantime, therefore, we have to rely on biochemical and bacteriological evidence which is in favour of xylitol.

Midda: Professor Frostell's last slide showed that these experiments should be repeated with mixtures of xylitol and/or other sugars to find whether there may be an ideal combination which will then perhaps have a cost benefit in view of the expense of xylitol.

Scheinin: In my opinion, Mr Chairman, the ideal commercial xylitol product does not yet exist. Today it should be a pure product—later on it might well be as suggested by Professor Frostell that certain mixtures might work but I would be extremely hesitant to consider mixtures of sucrose/xylitol to work in a desirable way at this time.

Midda: Well, we certainly seem to be a long way from the ultimate answer for the prevention of caries. Perhaps if we can come up with a sugar substitute that not only prevents caries but also periodontal diseases we will have the ultimate answer in dentistry. I think we agree that sugar is related to caries but all of us would agree that it is the frequency rather than the amount that is important and the real use of sugar substitutes is in the rôle of the in-between meal snack. Therefore, xylitol would appear to be an acceptable and promising material supported by biochemical, technological and clinical studies.

Index